Environmental Footprints and Eco-design of Products and Processes

Series editor

Subramanian Senthilkannan Muthu, SGS Hong Kong Limited,
Hong Kong, Hong Kong SAR

More information about this series at http://www.springer.com/series/13340

Andreja Kutnar · Subramanian Senthilkannan Muthu
Editors

Environmental Impacts of Traditional and Innovative Forest-based Bioproducts

Springer

Editors
Andreja Kutnar
University of Primorska
Koper
Slovenia

Subramanian Senthilkannan Muthu
Environmental Services Manager-Asia
SGS Hong Kong Limited
Hong Kong
Hong Kong

ISSN 2345-7651 ISSN 2345-766X (electronic)
Environmental Footprints and Eco-design of Products and Processes
ISBN 978-981-10-0653-1 ISBN 978-981-10-0655-5 (eBook)
DOI 10.1007/978-981-10-0655-5

Library of Congress Control Number: 2016932354

Printed on acid-free paper

This Springer imprint is published by Springer Nature
The registered company is Springer Science+Business Media Singapore Pte Ltd.

Preface

Forest biomass has been an important source of raw material for centuries. With the global focus on climate change and preservation of nonrenewable resources forest biomass has become an even more important resource. The world's political and economical decisions are increasingly determined by climate change and resource and energy scarcity. The forest-based industries have responded to this by developing advanced processes, materials, and wood-based solutions to meet evolving demands and increase competitiveness against non-forest bioproducts. This book provides a comprehensive description of traditional and innovative forest-based bioproducts. The descriptions of different types of forest-based bioproducts are followed with their environmental impacts from processing, use, and their end-of-life phase. Furthermore, the possibility of reusing, recycling, and upgrading bioproducts at the end of their life cycle is discussed. The book concludes with a discussion about the key issues and challenges for business development of forest-based bioproducts and about the business model innovation in the forest sector.

Andreja Kutnar
Subramanian Senthilkannan Muthu

Contents

Environmental Use of Wood Resources

Andreja Kutnar

Abstract This chapters delivers discussion about bioeconomy, related European Policy and wood resources. The European Bioeconomy Strategy and the role of forests and forest-products in achieving its objectives are being discussed. Also other policies with direct impacts on the forest-based sector are being presented. Furthermore, various primary wood based products and their environmental impacts are reviewed. The description covers traditional and innovative wood based products. The need for efficient resource use of wood resources, including reuse and upgrading of waste wood, are discussed. The environmentally preferred option to maintain wood materials in a maximum quality level by reuse in solid form, therefore extending the carbon storage duration, is presented.

Keywords Applications · Fibres · Manufacturing · Matrix components · Renewable composites · User perceptions · Wood plastic composites

1 Introduction

European Policy is affecting and, indeed, directing current research, development and marketing in the European Union (EU). Many policy strategies and actions directly impact the forest product industry. Europe is setting course for a resource-efficient and sustainable economy. The goal is a more innovative and low-emissions economy, reconciling demands for sustainable agriculture and fisheries, food security, and the sustainable use of renewable biological resources for industrial purposes, while ensuring biodiversity and environmental protection. The term being use to address this is bioeconomy. The bioeconomy comprises those parts of the economy that use renewable biological resources from land and sea—such as crops, forests, fish, animals and micro-organisms—to produce food, materials and

A. Kutnar (✉)
Faculty of Mathematics, Natural Sciences and Information Technology,
Andrej Marušič Institute, University of Primorska, Koper, Slovenia
e-mail: andreja.kutnar@upr.si

A. Kutnar and S.S. Muthu (eds.), *Environmental Impacts of Traditional and Innovative Forest-based Bioproducts*, Environmental Footprints and Eco-design of Products and Processes, DOI 10.1007/978-981-10-0655-5_1

energy. It is an essential alternative to the dangers and limitations of our current fossil-based economy and can be considered as the next wave in our economic development. It provides major opportunities for innovation, jobs and growth and as such will help to reindustrialise Europe. To achieve this, the European Commission has set a Bioeconomy Strategy and action plan. The strategy "Innovating for Sustainable Growth—A Bioeconomy for Europe" was launched and adopted on February 13, 2012 (European Commission 2012).

The European Bioeconomy Strategy focuses on three key aspects (European Commission 2012):

- Developing new technologies and processes for the bioeconomy;
- Developing markets and competitiveness in bioeconomy sectors;
- Pushing policymakers and stakeholders to work more closely together.

The Bioeconomy Strategy (European Commission 2012) emphasise the importance of forests, since the EU has a total forest area of approximately 177 million ha (around 40 % of the EU territory), of which 130 million ha are available for wood supply and the production of non-wood goods and services (cork, resins, berries, mushroom, hunting for example). The forest-based industries are a very important EU economic sector (woodworking industries, pulp and paper, printing industries), with a production value of € 365 billion, and an added value of around € 120 billion created by more than 3 million jobs (European Commission 2012).

Forests play a crucial role in the global carbon cycle and the fight against climate change. Forest biomass is currently the most important source of renewable energy and now accounts for around half of the EU's total renewable energy consumption (COM (2013) 659 final). The demand for wood, and for wood fuel in the context of increasing renewable energy demand, is a strong stimulus for increasing forest growth and productivity and for improving management practices more wood and residues could be harvested and mobilised while demand for forest products is growing for material and energy uses as a way to reduce carbon emissions by substituting products that cause higher emissions. Wood products can contribute to climate change mitigation as they act as a carbon pool during their service lives, as they withdraw CO_2 from its natural cycle, as wood products can substitute for more energy-intense products after their service life, as they can substitute for fossil fuels if they are incinerated (Werner et al. 2006).

Since increased harvest reduces carbon sinks, the Bioeconomy strategy (European Commission 2012) recognizes the need for speeding up production rates and developing forest raw materials with new properties. Forests of the future will be increasingly dedicated to producing fibres, timber, energy or customised needs, which will have considerable impacts on the provisioning of a broad range of public goods. However, it is important that forest products sector includes in its research and innovation activities also business development of forest based bioproducts. In this book Chap. 6 delivers the key issues and challenges for business development of forest based bioproducts. It defines the conditions and improvement potential for different product types: established forest products (building materials, paper and wood energy), potential large volume bioproducts (liquid biofuels), high value added products (biomaterials, and new chemicals, pharmaceuticals). It is illustrated that key challenges differ between 'old' bioproducts, such as wood where progress is connected with design development and incremental improvement of industrial processes—and 'radical' innovations of new materials and substances, which involves new market development. Furthermore, in Chap. 7 the product innovation and process innovation in forest sector companies are being compared. Chapter 7 explores the issues including consideration of pathways for forest sector firms to pursue in order to capitalize on the growing bioeconomy.

1.1 Policies with Direct Impacts on the Forest-Based Sector

Kuzman and Kutnar (2014) delivered an overview of the main policies with direct impacts on the forest-based sector. These policies are the EU Sustainable Development Strategy (SDS, European Commission 2009), which was published in 2006, and reviewed in 2009, the EU Roadmap 2050 (European Commission 2011), and the recycling society directive (Directive 2008/98/EC, European Parliament Council 2008).

The SDS sets out a single, coherent strategy on how the EU will more effectively live up to its long-standing commitment to meet the challenges of sustainable development. It recognizes the need to gradually change our current unsustainable consumption and production patterns and move towards a more integrated approach to policy-making. The overall intent of the SDS is to identify and develop actions to enable the EU to achieve continuous long-term improvement of quality of life. Specifically, the SDS calls for the creation of sustainable communities able to manage and use resources efficiently, able to tap the ecological and social innovation potential of the economy and in the end are able to enjoy prosperity, environmental protection and social cohesion.

The Roadmap 2050 project mission is to provide a practical, independent and objective analysis of pathways to achieve a low-carbon economy in Europe, which promotes energy security as well as the environmental and economic goals of the EU. Roadmap 2050 breaks new ground by outlining plausible ways to achieve an 80 % reduction in greenhouse gas emissions from a broad European perspective, based on the best available facts elicited from industry members and academia, and developed by a team of recognized experts rigorously applying established industry standards.

The latest waste directive from 2008 (Directive 2008/98/EC) contains an article for the re-use and recycling of all consumer and industrial materials. Amongst other things, it requires member countries to proceed with actions necessary to recycle materials as well as products. To fulfill these requirements, products should be developed with simple recycling as a product feature.

With the support of the EU Commission, industry stakeholders created the Forest-based Sector Technology Platform (FTP). FTP sees with its Strategic Research and Innovation Agenda for 2020 of the Forest-based Sector (Forest-based Sector Technology Platform 2013a) and the Horizons—Vision 2030 for the European Forest-based Sector (Forest-based Sector Technology Platform 2013b) this sector as a key actor and enabler of the biobased society. The FTP Vision 2030 (Forest-based Sector Technology Platform 2013b) is a strategy guide for the forest-based sector to help achieve the EU's goals of sustainable, inclusive growth.

1.2 Forest Technology Platform

The Strategic Research and Innovation Agenda (SRA) of the Forest-based Sector Technology Platform (2013a, b) defines 4 Strategic Themes and key research and innovation areas (RIAs):

- Strategic Theme 1: The forest-based sector in a biobased society
  - The performance of the sector in a perspective of global change
  - Citizens' perceptions of the sector and its products
  - Policies and good governance
- Strategic Theme 2: Responsible management of forest resources
  - Multi-purpose management of forests
  - Forest ecology and ecosystem services
  - Enhanced biomass production
  - Secured wood supply, forest operations and logistics
  - Cascade use, reuse and recycling systems
- Strategic Theme 3: Creating industrial leadership
  - Resource efficiency in manufacturing
  - Renewable energy solutions
  - Sustainable water stewardship
  - Biorefinery concepts
  - New business models and service concepts
- Strategic Theme 4: Fulfilling consumer needs
  - Building with wood
  - Indoor environment and functional furniture
  - New biobased products
  - Intelligent packaging solutions
  - Hygienic, diagnostic and healthcare products
  - Integration of new solutions in printed product

For each of the RIAs the SRA defines expected achievements by 2020 and required research and innovation activities. For example, for the cascade use, reuse, and recycling systems the expected achievement by 2020 is that the forest-based industry have developed and are following quality guidelines advocating a cascade use of wood along the following lines of priority:

1. Production of wood-based products
2. Re-use of products
3. Recycling into other wood-based products
4. Use as bioenergy source

To achieve this goal, the SRA defines for wood products the following requirements in the research and innovation activities:

- Improve the re-usability and recyclability of wood composites and construction material.
- Develop systems for wooden buildings allowing for easy dismantling and remounting.
- Develop environmentally-friendly additives and impregnating agents for wood products.
- Develop solutions for the utilisations of used wood from construction operations (scaffolds, concrete casting moulds) as a biorefinery raw material.

Innovative products optimized for end use requirements and sustainable resource utilization will lead to new business opportunities. This book addresses the required activities in the field od cascade use of wood and adds the required research and innovation activities in the field of building with wood and new business models and services. It is important that new concepts and methodologies to predict the market changes and consumer behavior are developed. Only with incorporating interdisciplinary knowledge and expertise new business models that target evolving consumer needs and behaviors can be developed. An important area addressing the consumer needs is building with wood. Since wood and wood products act as carbon sink and reduce CO_2 in the atmosphere, the use of structural wood material in place of concrete or other more emission-intensive construction materials is one of the most efficient ways of reducing the CO_2 emissions. The SGA (Forest-based Sector Technology Platform 2013a, b) foresees that until 2020 the wood-based construction methods will generally be perceived and credited as low carbon footprint construction, while the building regulations will have functionally-based requirements for product performance and will not discriminate against the use of wood in multi-storey constructions. Of course the activities related to responsible management of forest resources, including the multi-purpose management of forests. Wood as natural, renewable, reusable and recyclable raw material can play a major role in minimizing negative effects on the climate and environment, however only when it is sourced from sustainably-managed forests.

1.3 *Sustainable Forest Management*

In December 2007 the General Assembly of the United Nations adopted the most widely adopted intergovernmental definition of Sustainable Forest Management (SFM): "SFM as a dynamic and evolving concept aims to maintain and enhance the economic, social and environmental value of all types of forests, for the benefit of present and future generations." It is characterized by seven elements, including (UN 2008, Resolution 62/98):

(i) Extent of forest resources;
(ii) Forest biological diversity;
(iii) Forest health and vitality;
(iv) Productive functions of forest resources;
(v) Protective functions of forest resources;
(vi) Socio-economic functions of forests; and
(vii) Legal, policy and institutional framework

To promote the responsible management of forests and to prove that forests are well managed in accordance with strict environmental, social and economic criteria, forests have been certified to Forest Stewardship Council (FSC) and other certification schemes worldwide. FSC certification ensures SFM practices.

The FSC is a universal certification system, which was created in the early 90s in response to the pressing problem of deforestation, particularly in Africa. The basic objective of the FSC certification is to ensure tracking of FSC certified material from the forest to the consumer. The FSC Chain of Custody (CoC) system allows tracking FSC certified material from the forest to the consumer (Chain of Custody Certification 2015). It is a method by which companies can show their commitment to the environment and responsible forest management. Only companies that have FSC CoC certification are allowed to use the FSC trademarks and labels to promote their products. The FSC label therefore provides a link between responsible production and responsible consumption and helps the consumer to make socially and environmentally responsible buying decisions.

2 Wood-Based Products

Wood is in volume the most important renewable material resource (Rowell 2002). The utilization of wood in all aspects of human existence appears to be the most effective way to optimise the use of resources and to reduce the environmental impact associated with mankind's activities. However, as timber possesses good but not outstanding properties, this is not an easy thing to achieve, and in view of the new materials emerging it becomes noticeably more difficult. The only properties it has that reign supreme are ecological fitness and, possibly, a low cost.

Wood is a natural, renewable, reusable and recyclable raw material that can play a major role in minimizing the negative effects on the climate and environment when it is sourced from sustainably managed forests. The various species of wood have a number of physical characteristics that enable performance of wood needed in building construction. However, wood properties vary among species, between trees of the same species, and between pieces from the same tree. This leads to variability in the performance of wood, which is one of its inherent deficiencies as a material (Dinwoodie 2000). Therefore, a broad range of wood based composites has been developed in the past.

Below, the description of various primary wood based products is summarized and simplified from Suchsland (2004). Short description of glued laminated timber, Oriented Strand Board (OSB), Plywood, Cross-Laminated Timber (CLT), particleboard, and medium density fiberboard (MDF) is given.

Glued laminated timbers (Glulam) are structural composite beams used to support large loads in building construction (Fig. 1). Sawn timber, selected for stress-related mechanical properties, are glued and arranged in layers (with the high-grade timber in the outer layers, and low-grade timber in the inner layers) with the grain direction parallel to the length of the timber. The size of the resulting Glulam may vary greatly, allowing the beams to be used as needed for a specific application. Glulam for indoor use may use adhesives that are less resistant to the effects of the outdoor environment (e.g., relative humidity and temperature), while Glulam for outdoor use must use adhesives that are more resistant to changes in the outdoor environment.

OSB is a structural panel product most often used for roof, wall and floor sheathing in construction. The product is usually made of three or more layers with strands in each layer oriented in alternating directions (i.e., parallel to the length of the panel, or perpendicular to it). Water resistant adhesives are used for OSB. The strands in the outer layer are oriented with the grain direction parallel to the length of the panel. The strands used are typically about three times longer than they are wide.

Plywood is made from thin layers of wood, which has been peeled from a log on a rotary lathe. These thin veneers are then combined in three or more (usually an odd number) of layers in alternating grain directions. The outer layers are aligned with the grain direction parallel to the length of the panel. Plywood for indoor applications may use an adhesive that is less water resistant than plywood for outdoor use. In indoor applications plywood is often used in furniture. Plywood for outdoor applications must use a water-resistant adhesive. Sheathing is the most common use of plywood in exterior applications.

Beginning around 2000, CLT, a new type of wood-based panel, began development, and soon after went into commercial production. CLT is an engineered wood panel typically consisting of three, five, or seven layers of dimension lumber oriented at right angles to one another and then glued to form structural panels with exceptional strength, dimensional stability, and rigidity. An example of using CLT in complex structures is shown in Fig. 2.

Fig. 1 Use od glulam in "Pyramidentower" in Carinthia, Austria (Photo: Andreja Kutnar)

Wood composites produced in large quantities are particleboard and MDF. They are most commonly used for indoor, non-structural applications such as in furniture. Particleboard is constructed by reducing wood product manufacturing residues (e.g., planer shavings, sawdust) and recycled wood products to small particles. Particle sizes often vary across the thickness of the board, with smaller particles in the outer layers, and larger particles in the core layer. MDF is made by breaking wood (most often residues from other manufacturing processes) down to small

Fig. 2 CLT zurich elephant house (Photo: Andreja Kutnar)

fibers, then mixing the fibers with resin and wax to form mats that are compressed with pressure and heat.

Other more recent products include laminated veneer lumber (LVL), light MDF (LDF), and high density fibreboard (HDF). In the past years technological innovations have advanced the field of wood-based panels. Most notably, hot pressing and the consequent viability of thermosetting resins have improved composites produced from particles and strands (particleboard, OSB), fibers (as MDF, HDF) and veneers (plywood, LVL).

In spite of stronger regulations, the production of wood-based panels has recently experienced a dramatic, worldwide growth period. Europe and China each control more than 30 % of the worldwide capacity for wood-based panel production (Barbu and van Riet 2008). In Eastern Europe, new production is increasing, particularly in Commonwealth of Independent States (CIS) and Turkey. In Western Europe, Germany is the main wood-based panel producer (25 %), followed by France and Poland (10 % each) then Italy and Spain (8 % each). Turkey has dramatically increased production capacity and is now approaching Germany's capacity. Russia surpassed German production in 2011, but Germany may have latent capacity remaining from constricted production during the economic downturn (Kutnar and Burnard 2013).

Kutnar and Burnard (2013) presented past, present and future development of wood-based composites. A special emphasize is given to the trend to further restrict VOC and formaldehyde emissions, which requires innovative solutions and may increase the demand for biologically sourced adhesives as a preferred environmental and economic option in wood-based composites. In the last decade, wood-based panel production technology has adapted to the demands of ever changing markets with quality improvements, energy reductions, and more efficient raw material utilization. International research in this field has provided the industry with enhanced product technologies, which have been levied to improve product moisture resistance, mechanical and physical properties, reduce costs through energy savings (shorter pressing times) and more cost effective raw material utilization with cheaper and alternative raw materials, reuse and recycling (Carvalho 2008).

Environmental regulations and legislation regarding volatile organic compounds (VOCs) emissions (particularly with regards to formaldehyde in the wood products industry) are important driving forces for technological progress. Although panel product emissions have been dramatically reduced in recent decades, the reclassification of formaldehyde by the International Agency for Research and Cancer (IARC) as "carcinogenic to humans," is forcing panel manufacturers, adhesive suppliers and researchers to develop systems which further decrease formaldehyde emissions to levels as low as those present in natural wood (Athanassiadou et al. 2007).

2.1 Modified Wood and Thermo-Hydro-Mechanical Treatments

As a consequence of increased competition from traditional and new industries based on renewable resources, the forest resources must be considered to be limited. There are forecasts showing that, already in 2020, the European consumption of wood can be as large as the total European combined forest growth increment (Jonsson et al. 2011). Forest-based industries are continually developing advanced processes, materials and wood-based solutions to meet evolving demands and increase competitiveness.

One of the emerging treatments involves the combined use of temperature and moisture, where force can be applied, so that one speaks of thermo-hydro (TH) and thermo-hydro-mechanical (THM) processes. TH and THM processes play a vital role, as their point of application is decisive: resource efficiency and suitability for use. Kutnar et al. (2015) described thermally based timber processing as one emerging process technology which adds no toxic or other harmful substances. It can be divided into: TH treatments and THM treatments. A short historical overview of some well-known industrial processes based on TH or THM wood processing technology introduce the paper, indicating how the TH/THM technology has developed until today. These processes are at various stages of development, and the challenges that must be overcome in scaling up to industrial applications differ among them. To provide an understanding of the problems and possibilities related to the industrial implementation of the new TH/THM processes, the enhancement of product properties and the development of new wood products, some relevant scientific results are presented. Furthermore, the criteria within the framework of economic, environmental and social systems that must be met in the development of TH/THM processes to contribute to sustainable development and European low-carbon economy are discussed. Authors conclude that it is important to consider the effective use of wood throughout the value chain from forest management, through multiple use cycles, and end-of-life disposal.

Densification by THM process makes it possible for low-density and commercially uninteresting wood species to be modified into high performance and high value products. A recently developed method for wood densification using the viscoelastic thermal compression (VTC) process enabled the processing of thin materials (less than 10 mm) in the production of wood-base green composites (Kamke and Sizemore 2008; Kutnar et al. 2008). A simple application is a 3-layer laminated composite, with VTC wood in the two outer layers, and a layer of untreated low-density wood in the core (Fig. 3).

The THM processes and resulted products are discussed together with other modified wood products in Chap. 4. The low-carbon economy aims to mitigate climate change and promote sustainable development in Europe, by reducing energy consumption, pollution and emissions while increasing performance.

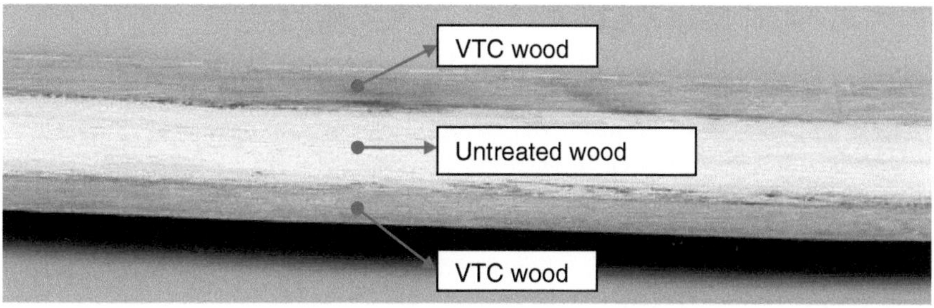

Fig. 3 The 3-layer VTC composite. The other laminas are 2.5 mm thick VTC wood laminas, whereas the core is 6 mm thick untreated wood

Therefore, research into timber processing and the resultant products must place more emphasis on the interactive assessment of process parameters, developed product properties, and environmental impacts. Energy consumption contributes considerably to the environmental impact of modified wood. However, the improved properties during the use phase might reduce the environmental impact of the timber processing. It is important to note that the effective use of wood throughout its whole value chain from forest management, through multiple use cycles, and end-of-life disposal can lead to a truly sustainable development.

2.2 *Environmental Impacts of Forest-Based Products*

Kutnar and Hill (2015) reviewed different methodologies that can be use to calculate the carbon footprint of wood products with the focus on their accounting for biogenic CO_2. Some methodologies, like Climate Declaration, do not account for biogenic CO_2. When using these methodologies, disregarding carbon storage in the assessment of the carbon footprint of wood products may bias the comparison with competing products that do not store biogenic carbon (Garcia and Freire 2014). To avoid misleading comparisons with other products, the embedded carbon should be reported when presenting cradle-to-gate results (Garcia and Freire 2014). Pawelzik et al. (2013) after comparing the methodology of the French Environment and Energy Management Agency (ADEME), the European Commission's Lead Market Initiative, the GHG Protocol Initiative, ISO/TS 14067, the International Reference Life Cycle Data (ILCD) Handbook, PAS 2050, and process/material carbon footprint, concluded that the lack of standardized approaches for accounting for the biogenic carbon storage in bio-based materials presents a key challenge to LCA calculations.

Hill et al. (2014) reviewed the methodologies for evaluating the atmospheric carbon stored in products. The ILCD Handbook, published by the European Commission Joint Research Centre (Institute for Environment and Sustainability) considers a 100 year assessment period and recommends that fossil and biogenic carbon releases (as carbon dioxide and methane) are differentiated. Carbon emissions associated with land use changes and from biomass associated with virgin forests are treated as fossil carbon. In contrast, emissions associated with plantation forests are inventoried as biogenic carbon. Uptake of atmospheric carbon dioxide is inventoried as 'resources from air'. A methodology is given for accounting for the removal and storage of atmospheric carbon dioxide. The uptake of atmospheric carbon dioxide is inventoried as 'Carbon Dioxide—Resources from Air' and the emissions as 'Carbon Dioxide (biogenic)—Emissions to Air'. These two flows then cancel each other out. Meanwhile, the issue of the storage in the product is calculated by declaring a correction flow for delayed emission of the carbon dioxide and giving it a value of 0.01 × the CO_2 equivalent mass stored per year. The same method is used to calculate the storage of fossil carbon in a long life product, except that there is no consideration given to the category 'Carbon Dioxide—Resources from Air'. Thus, there is a net effect of the release of the fossil derived CO_2 at the end of life, but the compensatory effect of the delayed emission of the fossil carbon is accounted for.

Kutnar and Hill (2014) used a cradle-to-gate analysis to present the carbon footprint of 14 different primary wood products. The largest source of emissions for all sawn timber products is removing the timber from the forest, while for kiln dried sawn timber the drying process follows closely behind. For fibre composites (MDF and HDF) the extra energy required to convert the raw material to fibres, in addition to the energy required to apply pressure and heat to the products is responsible for the bulk of the emissions from these products. The adhesives used in particleboard, plywood and OSB are responsible for the largest fraction of emissions from these products. This is especially significant considering the low total volume they represent in the final products. Glulam emissions derive mostly from the harvest and initial production of the softwood, but also from the extra energy required to apply pressure and set the adhesives used. Altering the system boundaries would yield different results. Furthermore, the results would have been different if the carbon footprint calculation accounted for carbon sequestration of wood, or considered the use of recycled wood products and other similar issues pertinent to Life Cycle Assessment (LCA).

In this book different wood-based products are presented. Chapter 2 reviews and discusses the performance and environmental impacts of wood plastic composites (WPCs) used in a variety of applications ranging from construction and automotive sectors to consumer goods. Performance is considered in terms of fitness for use, manufacturing methods, material components of WPCs, and user perceptions of the material. Additionally, the impacts of different additives being used in wood products are being discussed. The influence of these additives on service life, environmental implications, and what limitations they have on reuse, recycle and upgrading of wood products are being discussed in Chap. 4. Furthermore, in Chap. 5, the impact of wood-based insulation products on the environmental impacts of buildings are compared with the impacts of fossil-based insulation products.

2.3 *Reuse and Upgrading of Waste Wood*

Efficient resource use is the core concept of cascading, which is a sequential use of a certain resource for different purposes (Tavzes and Kutnar 2012; Höglmeier et al. 2013). This means that the same unit of a resource is used for multiple high-grade material applications (and therefore sequestering carbon for a greater duration) followed by a final use for energy generation and returning the stored carbon to the atmosphere. Intelligent concepts for reuse and recycling of valuable materials at the end of single product life will reduce the amount of waste to be landfilled.

Burnard et al. (2015) reviewed the European projects examining the recovered waste wood. The projects listed are DEMOWOOD (2015), CaReWood 2015, FPS COST Action E31 2011. According to these projects there is great potential to expand wood recovery for uses beyond energy and particleboard production (the two most common uses). Many solutions have been proposed to simplify and automate contaminant detection and then sorting and cleaning of the contaminated materials (DEMOWOOD 2013; Hasan et al. 2011). In addition to manual visual inspection and sorting, these solutions utilize a conglomerate of technologies. To detect and separate metals and other solid materials magnets, gravity sorting, rollers and sieves may be utilized (DEMOWOOD 2012). X-ray fluorescence systems (XRF), near infrared spectrometry (NIR), laser induced breakdown spectroscopy (LIBS), ion mobility spectrometry (IBS), and spectrally resolved thermography can be used to identify chemically treated wood waste (DEMOWOOD 2012). Once these technologies are implemented in a sorting facility, wood waste containing chemical compounds that are limited by law can be effectively identified and removed from processing. Chemically contaminated wood products of sufficient size may be resawn or planed to a smaller size, removing the contaminated (or otherwise damaged) surface(s) if the chemical penetration is not too deep. To re-enter the market, the newly produced timbers must be graded to certify their fitness for use. Removing physical contaminants such as nails or screws may require manual interventions, or sawing to remove heavily damaged or contaminated cross-sections (e.g., where large fasteners such as bolts, have left large holes).

In the wood-products sector, the waste hierarchy is presently underdeveloped and largely ignores the EU's preferred option of maximizing the carbon storage potential of wooden materials by their reuse in solid form, with subsequent downcycling of reclaimed wood in as many steps of a material cascade as possible (Leek 2010). At present, in Europe recovered wood volumes total approximately 55.4 million m^3. One third of this volume is burned for energy production, and one third is downcycled and used for the production of particleboard, thus losing the favorable material properties of solid wood. The remaining (and largest) fraction of waste wood (20.4 million m^3) is not currently used within the EU27 and is landfilled (Leek 2010). However, this ignores the environmentally preferred option to maintain wood materials in a maximum quality level by reuse in solid form, therefore extending the carbon storage duration. This shortfall presents an opportunity for the forest-based sector to become a leader in achieving the European Commission's ambitious target of reduced CO_2 emissions with innovative production technologies, reduced energy consumption, increased wood products' recycling, and the reuse and refining of side-streams (e.g., manufacturing byproducts such as sawdust as planer shavings). However, innovations going beyond the wood-products will have to follow. It is important that new business models are developed by taking into account the reverse logistics.

Burnard et al. (2015) presented a case study of reverse logistics of waste wood and wood products. It includes the coordination and control, physical pickup and delivery of the material, parts, and products from the field to processing and recycling or disposal, and subsequent returns back to the field, where appropriate. The authors provide descriptions of the services related to receiving the returns from the field, and the processes required to diagnose, evaluate, repair, and/or dispose of the returned units, products, parts, subassemblies, and material, either back to the direct/forward supply chain or into secondary markets or full disposal. The developed model does not include a return on investment calculation for placed facilities and legal parameters that define new categories and quality assessments in each of the entities.

In this book the use of recovered construction wood is discussed in Chap. 3. The opportunities and challenges including the factors which influence the re-use and recycling of wood are presented.

3 Conclusion

In order to attain a sustainable and healthy society, a shift towards sustainable activities, consumption, production and innovation is necessary. Despite many sustainable and green building activities in recent years there is a significant space for expansion and improvement in this area. In fact, a shift towards new building paradigms is just beginning. This change aims at minimal and zero environmental loading, towards positive impacts for both the environment and society. This does not necessarily mean the zero-energy or negative-energy buildings but seeks the balance between environmental impact, societal needs, available technologies and affordability to enhance sustainability. The buildings of new generation must be environmentally friendly and at the same time affordable.

The need to reduce the whole-life energy consumption of buildings has highlighted the role that wood can play in construction. Wood is a good thermal insulator making it an ideal material for energy-efficient construction. When buildings have net zero energy consumption, a major part their overall environmental burden consists of their embodied energy and the associated greenhouse gas emissions. Compared with other construction materials such as steel or concrete, the energy needed to convert a tree into the final product is significantly lower, resulting in wood products having a low embodied energy. However, a renewable origin does not necessarily mean 'environmentally friendly' or sustainable use. Wood harvest, manufacturing and its use produce far less greenhouse gas emissions than other materials. Fossil fuel consumption, potential contributions to the greenhouse effect and quantities of solid waste tend to be minor for wood products compared to competing products. Impregnated wood products tend to be more critical than comparative products with respect to toxicological effects and/or photo-generated smog depending on the type of preservative. Furthermore, although composite wood products such as particleboard and fibreboard make use of a larger share of the wood of a tree compared to products out of solid wood, there is a high consumption of fossil energy associated with the production of fibres and particles/chips as well as with the production of glues, resins, etc. Also, treated wood, adhesively bonded wood and coated wood, might have toxicological impacts on human health and ecosystems. Despite of this renewably-sourced materials are expected to play a vital role in bioeconomy because:

- Materials that reflect the natural environment will motivate users to care for their environment and cause a shift in consumption patterns towards sustainably sourced and produced materials.
- The demand for renewable-sourced materials will drive innovation in the forest-based sector, which will support more jobs and strengthen the economy.
- The forest-based sector offers many opportunities for new products, processes and systems that only well supported and executed research can identify and create value from.

The forest products can be use for a larger number of application. The demand for more environmentally sustainable products and energy is making the wood resources an important part of the bioeconomy.

Acknowledgments The author is pleased to acknowledge the support of WoodWisdom-Net + and the Slovenian Ministry of Education, Science, and Sport of the Republic of Slovenia for their support of the What We Wood Believe and Cascading Recovered Wood projects; European Commission for funding the project InnoRenew CoE under the Horizon2020 Widespread-2015 program, and infrastructure program IP-0035.

References

Athanassiadou E, Tsiantzi S, Markessini C (2007) Towards composites with formaldehyde emission at natural wood levels. In: Proceedings of the 2nd conference of cost action E49, Measurement and control of VOC emissions from wood-based panels, Braunschweig, GER, pp 28–29

Barbu M, Van Riet C (2008) European panels market developments—Current situation and trends. In: Proceedings of the SWST annual convention, Concepción, Chile, Society of wood science and technology, Madison, WI

Burnard DM, Tavzes Č, Tošić A, Brodnik A, Kutnar A (2015) The role of reverse logistics in recycling of wood products. In: Muthu SS (ed) Environmental implications of recycling and recycled products, Environmental footprints and eco-design of products and processes, ISSN 2345–7651, Springer, Singapore

Carvalho LM (2008) Modelling of the hot-pressing process. In: Core document of COST E34 "Bonding Timber", Lignovisionen Issue 18, Wien, AUT: Universität für Bodenkultur

Chain of custody certification (2015) https://ic.fsc.org/chain-of-custody-certification.39.htm. Accessed 24 Dec 2015

DEMOWOOD (2012) Optimisation of material recycling and energy recovery from waste and demolition wood in different value chains NWP2-ER-2009-235066, Deliverable for DL—WP2.1. http://www.wwnet-demowood.eu/fileadmin/PTS/Demowood/Dokumente/DL_WP2.1_Sorting%20Techniques_pts_120607.pdf. Accessed 25 Dec 2015

DEMOWOOD (2013) Optimisation of material recycling and energy recovery from waste and demolition wood in different value chains NWP2-ER-2009-235066, Deliverable for DL—WP2.2. http://www.wwnet-demowood.eu/fileadmin/PTS/Demowood/Dokumente/DL_WP2.2_Quality%20Assessment%20of%20Waste%20Wood.pdf Accessed 25 Dec 2015

DEMOWOOD (2015) Material recycling and energy recovery from waste and demolition wood. http://www.wwnet-demowood.eu/index.php?id=1354. Accessed 24 Dec 2015

Dinwoodie JM (2000) Timber: Its nature and behaviour, 2nd edn. EFN Spon, London, New York, p 257

European Commission (2009) Mainstreaming sustainable development into EU policies: 2009 Review of the EU strategy for sustainable development, Communication. Brussels: European Commission. http://eur-lex.europa.eu/LexUriServ/LexUriServ.do?uri=CELEX:52009DC0400:EN:NOT. Accessed 15 Dec 2015

European Commission (2011) A Roadmap for moving to a competitive low carbon economy in 2050, Communication. Brussels: European Commission. http://eur-lex.europa.eu/LexUriServ/LexUriServ.do?uri=CELEX:52011DC0112:EN:NOT. Accessed 15 Dec 2015

European Commission (2012) Innovating for sustainable growth—A bioeconomy for Europe, Luxembourg, Publication office of the EU

European Parliament Council (2008) Directive 2008/98/EC of the European Parliament and of the council of 19 November 2008 on waste and repealing certain Directives, Directive. Brussels: European Parliament. http://eur-lex.europa.eu/LexUriServ/LexUriServ.do?uri=CELEX:32008L0098:EN:NOT. Accessed 15 Dec 2015

Forest-based Sector Technology Platform (2013a) Horizons—Vision 2030 for the European Forest-based sector, Research Agenda. Brussels: Filip De Jaeger, Gérant FTP Forest-based sector technology platform

Forest-based Sector Technology Platform (2013b) Strategic research and innovation agenda for 2020, Research Agenda. Brussels: Filip de Jaeger, Gérant FTP Forest-based Sector Technology Platform

Garcia R, Freire F (2014) Carbon footprint of particleboard: A comparison between ISO/TS 14067, GHG protocol, PAS 2050 and Climate declaration. J Clean Prod 66:199–209

Hasan AR, Schindler J, Solo-Gabriele HM, Townsend TG (2011) Online sorting of recovered wood waste by automated XRF-technology. Part I: Detection of preservative-treated wood waste. Waste Manage 31(4):688–694

Hill C, Norton A, Kutnar A (2014) Environmental impacts of wood composites and legislative obligations. In: Ansell M (ed) Wood composites—From nanocellulose to superstructures, Woodhead, Cambridge, UK

Höglmeier K, Weber-Blaschke G, Richter K (2013) Potentials for cascading of recovered wood from building deconstruction—A case study for south-east Germany. Resour Conser Recy 78:81–91

Jonsson R, Egnell G, Baudin A (2011) Swedish forest sector outlook study. United Nations, Geneva

Kamke FA, Sizemore H (2008) Viscoelastic thermal compression of wood, U.S. Patent No. 7,404,422

Kutnar A, Burnard MD (2013) The past, present, and future of EU wood adhesive research and market. In: International conference on wood adhesives, Forest products society, Toronto, Ontario, Canada, Madison, 9–11 Oct 2013

Kutnar A, Hill C (2014) Assessment of carbon footprinting in the wood industry. In: Muthu SS (ed) Assessment of carbon footprint in different industrial sectors, EcoProduction, Vol 2. Springer, Singapore, pp 135–172

Kutnar A, Hill C (2015) End of life scenarios and the carbon footprint of wood cladding. In: Muthu SS (ed) The carbon footprint handbook, CRC Press, Taylor & Francis Group, New York

Kutnar A, Kamke FA, Sernek M (2008) The mechanical properties of densified VTC wood relevant for structural composites. Holz als Roh- und Werkstoff 66(6):439–446

Kutnar A, Sandber D, Haller P (2015) Compressed and moulded wood from processing to products—A review. Holzforschung 69(7):885–897

Kuzman MK, Kutnar A (2014) Contemporary slovenian timber architecture for sustainbility. Green energy and technology, ISSN 1865–3529, Springer, Switzerland

Leek N (2010) Post-consumer wood. In: EUwood—Real potential for changes in growth and use of EU forests, Final Report, EUwood, Hamburg, Germany

Pawelzik P, Carus M, Hotchkiss J, Narayan R, Selke S, Wellisch M, Weiss M, Wicke B, Patel MK (2013) Critical aspects in the LCA of bio-based materials—Reviewing methodologies and deriving recommendations. Resour Conserv Recycl 73:211–228

Rowell RM (2002) Sustainable composites from natural resources. In: Brebbia CA, de Wilde WP (eds) High performance structures and composites. WIT Press Boston, MA, pp 183–192

Suchsland O (2004) The swelling and shrinking of wood: A practical technology primer. Forest Products Society, Madison, Wisconsion, USA

Tavzes Č, Kutnar A (2012) Koncept "kaskade" ali "spirale" uporabe lesa. In: Kitek Kuzman M. Lesene konstrukcije v stanovanjski in javni gradnji: Slovenija. Ljubljana, Biotehniška fakulteta, Oddelek za lesarstvo, Fakulteta za arhitekturo: 38

UN (2008) Non-legally binding instrument on all types of forests. UN Resolution A/RES/62/98, New York, USA

Werner F, Taverna R, Hofer P, Richter K (2006) Greenhouse gas dynamics of an increased use of wood in buildings in Switzerland. Clim Chang 74:319–347

Wood-Plastic Composites—Performance and Environmental Impacts

Matthew John Schwarzkopf and Michael David Burnard

Abstract This chapter reviews and discusses the performance and environmental impacts of wood-plastic composites (WPCs) used in a variety of applications ranging from construction and automotive sectors to consumer goods. Performance is considered in terms of fitness for use, manufacturing methods, material components of WPCs, and user perceptions of the material. Recent research related to matrix components and their relation to mechanical properties are covered in detail, especially regarding effects of the wood component. Manufacturing processes are also significant contributors to the suitability of WPCs for a given use, and the impact of various aspects of manufacturing are discussed as well. The environmental impacts of WPCs are reviewed and contain comparisons to solid wood alternatives, different matrix components, and future considerations for performing environmental impact assessments of WPCs. Finally, critical aspects of further innovation and future research are covered that are necessary to improve WPCs use as suitable replacements for solid plastic products and materials.

Keywords Applications · Fibres · Manufacturing · Matrix components · Renewable composites · User perceptions · Wood-plastic composites

1 Introduction

Wood-plastic composites (WPCs) are a product class that has been developing over the last 40 years resulting in increased applications and expanded market share. More specifically, WPCs are composites containing a wood component in particle form (wood particles/wood flour) and a polymer matrix. They are used in a variety

M.J. Schwarzkopf (✉) · M.D. Burnard
Andrej Marušič Institute, University of Primorska,
Muzejski Trg 2, Koper, Slovenia 6000, Slovenia
e-mail: matthew.schwarzkopf@iam.upr.si

M.D. Burnard
Research and Development, Abelium, d.o.o, Ljubljana, Slovenia

A. Kutnar and S.S. Muthu (eds.), *Environmental Impacts of Traditional and Innovative Forest-based Bioproducts*, Environmental Footprints and Eco-design of Products and Processes, DOI 10.1007/978-981-10-0655-5_2

of structural and non-structural applications ranging from component and product prototyping to outdoor decking. However, construction and automotive applications are the most common worldwide (La Mantia and Morreale 2011; Eder and Carus 2013). WPCs can be used outdoors as well as indoors, and some common applications include construction materials, garden and yard products, automotive applications (interior and engine), household items, packaging and consumer goods.

The decision to use a WPC product in place of another, generally speaking, should be predicated on achieving greater performance, reduced price, or reduced environmental impact. Using exterior decking as an example of improved performance, a homeowner may choose to use WPC decking instead of pressure-treated timber because of the ease of maintenance and improved durability (i.e. coatings do not need to be applied), or aesthetic reasons (e.g. lack of knots and splits likely to be found in solid wood).

The benefits are not always clear though, as there are many trade-offs to consider when choosing WPCs—especially when replacing responsibly sourced renewable materials. Including any non-renewable materials in the product can significantly increase its environmental impact. However, the case for WPCs may be clearer when the competing product is made entirely of non-renewable polymers, as the wood fraction of many WPCs can approach 50 % of the product volume, and therefore reduce the resource pressure on non-renewable materials (e.g. polymers derived from fossil sources). The environmental impact of WPCs is directly affected by the ratio of renewable to non-renewable materials in the product. Not only do WPCs have lower environmental impacts than unfilled plastics (but higher than solid wood or most other wood composites), use of sustainably harvested and recovered wood products in long-life products sequesters atmospheric carbon and can produce a positive environmental impact (Hill et al. 2015).

Wood is often used in plastics as a means to reduce price compared to a solid plastic product. Wood used in WPCs often comes from side streams such as sawdust produced while manufacturing lumber or recovered wood products, and is much cheaper to produce than the plastic that it replaces in many products. This often helps to reduce prices for consumers.

Promising progress and research into bioplastics [i.e. plastics made from biopolymers such as Polylactic acid (PLA)] reinforced with natural fibres (including wood) indicate the potential for these renewable materials to eventually enter the WPC market (Mukherjee and Kao 2011). However, these composites still have significant environmental impact due to processing and production, and steps to reduce the energy demands and water use should be taken (Qiang et al. 2014).

WPCs already have a significant market share (260,000 t in Europe in 2012) and the trend is increasing (Eder and Carus 2013). The current reliance on plastics, especially those derived from fossil sources, means that this demand is likely to continue increasing especially as developing economies continue to grow. However, to meet industrial, consumer, and environmental demand's research must continue to improve the processes for making WPCS, the component materials, and final product functionality.

This chapter will give an overview of the material properties, manufacturing processes, applications, and current developments related to WPCs.

2 Components of WPCs

WPCs are a group of composite materials and products comprised of two primary and distinct phases. One of these phases is the matrix which holds the different components together, binding them and providing load transfer between them. The matrix in WPCs is either a thermoset or more commonly a thermoplastic polymer. The other primary phase is the wood component. The wood component can be of any shape or size and acts as a filler and/or reinforcement to the composite. Making up a relatively small proportion of the total composite are additives which are added to aid in processing and affect a variety properties of the final product.

2.1 Matrix Component

2.1.1 Thermosets

Thermosets are a class of polymers that upon curing cannot be remelted or reprocessed for the same type of usage. From a liquid state these polymers are cured into rigid solids that are chemically cross-linked. The mechanical properties of these polymers come from initial molecular units and the density of cross-links formed during curing (Hull and Clyne 1996). The wood adhesives industry often uses thermosets to take advantage of this cross-linked form which provides a solid and durable bond. When used with wood, the liquid polymer penetrates the wood microstructure to varying degrees and is then cured, forming a three dimensionally dispersed interphase region. Common thermosets being used are urea–formaldehyde, phenol–formaldehyde, epoxy, and polyamides. One of the first WPCs using a thermosetting polymer was a phenol–formaldehyde—wood composite which is branded Bakelite by Rolls Royce in 1916, used for the shifting knob in their vehicles (Clemons 2002).

2.1.2 Thermoplastics

Thermoplastics are a class of polymers that can be heated and softened, cooled and hardened, and then resoftened while maintaining their characteristic properties from their first usage. Thermoplastics are used for a variety of everyday products like plastic soda bottles, single use shopping bags, milk jugs, etc. Unlike thermosets, these polymers are not cross-linked and rely on the properties of their monomer units, large molecular weights, and polymer chain entanglement for their mechanical performance (Hull and Clyne 1996). When heated, these polymer chains disentangle and allow them to slide past each other, allowing for reprocessing. High density polyethylene (HDPE), polypropylene (PP) and polyvinyl chloride (PVC) are the most common thermoplastic polymers used in WPCs (Klyosov 2007). HDPE accounts for the majority of the thermoplastics used in WPCs at 83 %, followed by PP at 9 % and PVC making up 7 % of the total WPC thermoplastic volume (Caulfield et al. 2005). Due to the thermal stability of wood, thermoplastics are used because they can be processed at relatively low temperatures below wood's thermal degradation temperature (180–200 °C). These polymers are also attractive for WPCs because they can be cut, screwed, and nailed with tools already used for wood construction.

2.2 Wood Component

Polymer manufacturers have historically used minerals and synthetic materials like talc, calcium carbonate, mica, glass fibres, and carbon fibres as extenders and reinforcing materials (Eckert 2000). Wood as a filler or reinforcing material has been used in composite materials for thousands of years (Bodig and Jayne 1982) and the introduction of a natural filler like wood particles in polymers was appealing to polymer manufacturers. Wood has many advantages to traditional fillers like lower cost, relatively high strength to weight ratio, low density, is relatively soft and easily integrated into existing plastic production lines, can offset the amount of polymer used, and is a renewable resource (Wolcott and Englund 1999; Clemons 2002; Farsi 2012). English et al. (1996) found that wood flour used in PP composites offers similar performance to that of talc and other mineral fillers but with a lower specific gravity providing for lighter composites.

In its own right, wood is a naturally occurring composite utilising polymers in a highly structured cellular construction. A thorough treatment of wood and its constituents related to composite materials can be found in Bodig and Jayne (1982). The wood material used in WPCs can be from a virgin source or is often a post-industrial co-product like trimmings from sawmills, breakdown of urban and demolition wood, or logging trimmings/slash. These materials are then chipped and ground into their final form as wood particles. Typical methods for the comminution of wood into wood particles are through the use of hammer and attrition mills. Unlike clear (free of defects) and undamaged wood, the wood particles often used in WPCs have been heavily altered. Using this type of mechanical breakdown, the finished product (Fig. 1) is heavily damaged and its properties are far from those of defect free, clear wood.

Particles are typically less than 1 mm in length and have a wide distribution of aspect ratios (Wang 2007). These particles are comprised of bundles of short fibres rather than long individual wood fibres. The complex morphology of the particle cellular structure, its irregular geometry, and the damage caused through processing should be acknowledged when designing and manufacturing WPCs (Teuber et al.

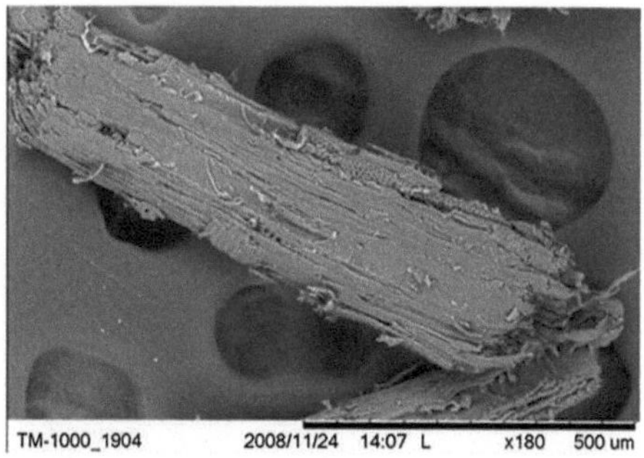

Fig. 1 SEM image of a wood particle used in WPCs (Schwarzkopf 2014)

2015). WPC properties are significantly influenced by the wood species and particle size characteristics of the wood particles being used (Stark and Rowlands 2003).

2.2.1 Wood Species

The wood species used in WPCs is typically determined by geographical location, availability, and price. The wood species used affects important aspects of WPC production like chemical compatibility and mechanical contributions to the composite. Common wood species used are pine, maple, and oak. Berger and Stark (1997) tested a variety of wood species in injected-moulded WPCs. They found that hardwood species provide improved tensile properties and heat deflection when compared with softwoods, and that ponderosa pine wood flour provided the optimum blend of mechanical property enhancements. These results are by no means the answer for all situations but show that wood species selection is an important factor to consider.

One aspect dependent on the wood species selection is the microstructure of the wood itself. The effective surface area for interaction with the polymer and the degree of polymer penetration into the wood structure both affect the composite properties. Escobar and Wolcott (2008) investigated the influence of different species on WPCs. Part of this study looked at the effect that different anatomical features of wood had on the polymer penetration through the wood structure. For example, Fig. 2 shows a cross-sectional view of a wood specimen with cell lumens filled with HDPE.

Gacitua and Wolcott (2009) found that wood species with higher interfacial areas (Fig. 2) may increase the amount of mechanical interlocking of the polymer with the wood structure. Circled in yellow are examples of the interfacial areas where the wood and polymer interact. This area also includes the entire perimeter of the wood particle. Depending on which wood species is used, the microstructure may be much different and increase or decrease this interaction area. It is also important to note that the degree of polymer penetration within the wood structure is also affected by the polymer's composition with respect to molecular weights. Low molecular weight components can more easily penetrate the wood structure, but contribute less to mechanical properties and an optimum balance must be found.

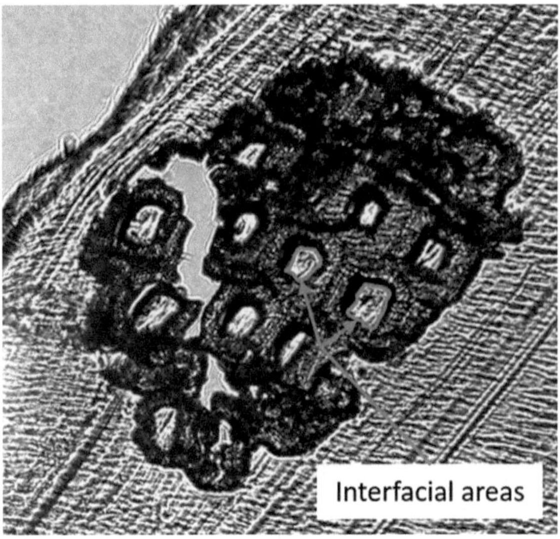

Fig. 2 Wood particle embedded in PE showing cell lumen penetration and highlighting areas making up the entire wood–polymer interaction area. *Photo credit* Muszyński, L

2.2.2 Wood Particle Size

The size and geometry of the wood particles being used in WPCs affect the flow/handling characteristics and mechanical properties. Wood particles obtained on an industrial scale often have a large-size distribution (Fig. 3) making it more difficult to design for certain properties. However, the more milling and screening that are done to make the particles smaller or narrow this distribution increase the cost

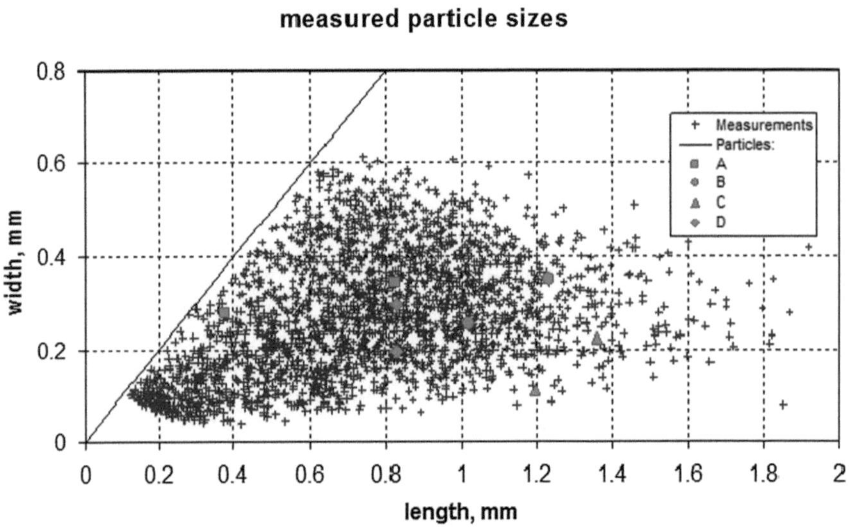

Fig. 3 Particle size distribution scatter plot from a 40-mesh sample (Wang 2007)

significantly. Particle size is characterised by the size of spaces in a mesh screen the particles pass through. Mesh size refers to how many openings there are in a screen within 2.54 cm^2. For example, a 100-mesh screen has 100 openings in 2.54 cm^2. The higher the mesh number, the smaller the particles. Mesh sizes of particles used in WPCs will vary depending upon the desired product properties and finish, and are most commonly from 10 to 80 mesh (Patterson 2001; Clemons 2002).

Stark and Rowlands (2003) investigated the effect of particle size on the mechanical properties of wood/PP composites. They manufactured WPC test specimens using particle mesh sizes 35–235 and performed a variety of mechanical tests. They found WPCs with larger particle sizes contained more stress concentrations which affected the impact energy of the product. They also found that even more important than particle size was the aspect ratio of the particles which is the length divided by the width of the largest minor axis of the particles. Generally, when there are particles with a larger aspect ratio, there is the potential for more effective load transfer between the matrix and the particles leading to better mechanical properties (Schwarzkopf and Muszynski 2015). Based on a 40-mesh sample of commercially obtained wood particles, Wang (2007) investigated the distribution of particle sizes. Using optical measurements of micrographs, Wang found that the median aspect ratio was 2.8. Using fillers with greater aspect ratios like wood fibre (10–20) can improve mechanical properties of the WPC, but difficulties in processing occur when feeding and metering the fibres into extruders (Patterson 2001).

3 Manufacturing Methods

WPCs started being produced by the plastics industry which had prior expertise in processing and manufacturing of plastic products (Clemons 2002). This industry had used filler materials in the past and when wood became a viable option, it was integrated into their existing production lines. While other wood-based composites are typically made in a panel or beam like geometry. WPCs starting in a molten state can be formed into highly detailed, linear profiles using extrusion processes or can be formed into complicated shapes via injection moulding. In any thermoplastic composite, the components must first be blended together and then later formed into the desired product.

3.1 Compounding

Mixing or compounding is the act of combining the wood and polymer components together. During the compounding procedure it is critical to evenly disperse the wood particles throughout the molten polymer. This dispersion is especially important with highly filled WPCs (Schirp and Stender 2009). It is also important in this step to wet or encapsulate the wood particles with the polymer. Proper dispersion and wetting allow uniform and more effective load transfer to occur throughout the composite. If not compounded properly, the composite will have reduced mechanical properties compared with an optimally compounded blend and increases the risk of durability issues. After compounding, the material can go directly to shape formation of the final product or can be chipped into pellets for later use.

3.2 Extrusion

The majority of WPCs are extruded into long linear profiles to use as decking planks, siding, fences, etc. Extruders serve the two main purposes of compounding the wood and filler, and then forming the shape of the extruded profile.

The wood and polymer components are metered and fed into the extruder and mixed using single or twin-screw configurations. The screws act to mix and move the material forward. Throughout the barrel of the extruder, the mix is heated through friction between the barrel, screw, and wood–polymer mix as well as by heated zones along the length. At the end of the extruder is a die through which the material is fed, forming the desired profile. Twin screw extruders are sometimes used as compounding units for producing pre-blended pellets. Manufacturers using a single screw extruder or injection moulding process often purchase these pre-blended pellets which are more easily fed into the machine and do not require an extra compounding step.

3.3 Injection Moulding

Injection moulding is used much less for WPCs, but can be used to make more complex shapes for a variety of products. The first steps in injection moulding are similar to extrusion, but instead of being forced through a die, the mixed material is injected into a mould. The wood-plastic mixture fills the mould, is cooled, and is then ejected in the preparation for the next piece to be formed.

3.4 Wet Processes for Sheet Formation

Sheets of WPC, which are often used in the automotive industry (e.g. doors or shelving applications), are either extruded or formed by a wet process. In wet process fabrication, a slurry of water and wood is created and mixed with chemical additives before being hot pressed into sheets (Pritchard 2004). These sheets may use a plastic scrim to help holding the board together (Pritchard 2004).

3.4.1 3D Modelling with WPCs

3D modelling (or additive manufacturing) is a manufacturing technique that allows for complex shapes to be created by depositing or removing materials (such as WPCs and other plastics, but a variety of materials can be used) in a customisable pattern to make three-dimensional objects. For most of these processes object models are created in a 3D modelling environment (such as computed-aided drafting (CAD) software), then processed in software that splits the 3D object into a collection of layered elements which can be created by the modelling technique.

3D modelling is most frequently used as a rapid prototyping method. Prototyping allows the users to create and test variations of product designs. Commercial applications for 3D modelling are expanding in a range of fields, however. Furthermore, the affordability of non-commercial 3D printers (particularly Fused Deposition Modelling (FDM) systems) has allowed researchers, hobbyists, and small scale component manufacturers to explore a variety of materials, methods, and products. WPCs are used in 3D modelling to reduce material costs and reduce the environmental impact using fossil-based plastics. As in other WPC applications, waxes, photostabilisers, lubricants, and other additives are used in the material matrix to alter the properties of the final product and aid in the manufacturing process.

FDM is a leading 3D modelling method in many manufacturing areas (Nikzad et al. 2011). The WPC used in this method is a filament that is fed into a nozzle that heats and deposits the WPC according to the product design. The WPC must be heated to a pliable state without exceeding the thermal degradation temperature of the wood fibre in the filament, which can limit the types of plastics used in these applications.

Selective laser sintering (SLS) is another 3D modelling technique which is in developmental stages for use with WPCs (Guo et al. 2011). SLS utilises powders which melt at different temperatures and that are fused together by laser radiation and form solids as the temperature of the combined material decreases. The methods for preparing WPCs for SLS are underdevelopment, but the wood component must be treated (alkalised) and mixed with a thermoplastic adhesive powder.

3.5 *Reinforcement of Plastic Matrices with Renewable Materials*

The primary purpose of using renewable fibre reinforcement in plastics is to reduce material cost, which has a secondary effect of reducing ecological impacts, especially when replacing non-renewable reinforcement (e.g. metals and glass) (Corbière-Nicollier et al. 2001). However, reinforcing plastics with particulates and fibres impacts material properties (strength, durability, appearance, etc.) as well as their ecological impact (Corbière-Nicollier et al. 2001; Zhong et al. 2001; Bouafif et al. 2009; Westman et al. 2010; Mukherjee and Kao 2011). Using renewable fibres and particulates alter the properties of the composite material in a variety of ways based on the geometry of the reinforcement, the type and components of the matrix the renewable components are part of, and the type of fibre or particle embedded in the matrix (Mukherjee and Kao 2011). Materials such as wood, reeds, kenaf, grasses (like bamboo), cotton, carbon fibres, rayon, nylons, and many other renewables allow reduced demand on fossil and other non-renewable (e.g. metals and glass) matrix components.

4 Physical Characteristics

Composite materials are often optimised by selecting components for their strength, stiffness, flexibility, and durability. When compared with individual materials, composites may also offer more consistent performance, lower production costs, and create an avenue for the utilisation of renewable resources. WPCs are no different and are formulated to meet the needs of the consumer by finding the right balance of these properties. Mechanical properties and durability are among the most important to WPCs.

4.1 Mechanical Properties

With WPC decking making up the largest share of the WPC market (Clemons 2002), we can look at mechanical properties important to this market. WPC deck boards are subjected to bending when they span a gap between supports and are being dynamically loaded when walked on and supporting the static loads (e.g. furniture and grills). Both the ultimate tensile stress (UTS) and modulus of elasticity (MOE) are important properties to consider. UTS is the maximum stress that a material can be subjected to before breaking. MOE refers to a material's ability to resist deformation and in a general sense is the stiffness of the material. For decking, this is important for limiting deflection of the product. It should be mentioned that a true elastic response in plastic composites is debatable, and the response of the material is highly dependent on the testing rate, temperature, previous history of the specimen, etc. Comparing values between different profiles, WPC formulations and specimens from different testing facilities is difficult, but for research and development purposes determining these values as a comparison is helpful. A study done by Karas (2010), assessed a variety of mechanical properties including MOE and UTS for wood-HDPE composites. Commercial pine flour was used as a filler as well as a variety of wood fillers from "low-grade" sources including: whole-tree juniper (WJ) (including bark), forest thinning material (FT), and urban wood from demolition (UW).

In the left plot in Fig. 4, UTS is plotted against the filler loading ratio. There is a horizontal line at 20 MPa which represents the UTS value for the HDPE used in this experiment that contains no filler. The other lines represent composites made from HDPE and the various filler types mentioned above. When increasing the wood filler loading ratio there is a slight increase in UTS, but with higher loading levels near 60 % wood, the UTS decreases. This behaviour is expected because when more and more of the composite is wood, the particles are often not entirely encapsulated by the polymer and optimal load transfer is not possible. The right plot in Fig. 4 is showing the MOE plotted against the loading ratio of wood fillers. As the filler ratio increases from 0 to 60 %, the MOE increases for all of the samples. This stiffening behaviour is also present in composites using fillers other than wood. Filling WPCs above 60 % requires care in particle dispersion and increases the likelihood of problems with not fully encapsulated particles, water absorption, crack formation, and biological attack.

WPCs have found success in a variety of markets including outdoor decking, railings, fences and landscaping timbers, but the number of applications for WPCs is limited to service not requiring high-mechanical performance (Clemons 2002). Hull and Clyne (1996) stated that understanding load transfer is the key to understanding the composite's mechanical behaviour. In WPCs, commonly used thermoplastic polymers like PP are hydrophobic (water-hating) while the constituent polymers of wood, like cellulose, are hydrophilic (water loving) in nature and have reactive hydroxyl groups along the length of their chains (Sjöström 1993). This results in an incompatibility between the polar wood component and the non-polar thermoplastic materials resulting in poor adhesion between the two (Lu et al. 2000) and lower mechanical properties than properly bonded components.

Attempts to improve the quality of these bonds have been made in the past by experimenting with additives known as coupling agents. Coupling agents are defined by Pritchard (1998) as "substances that are used in small quantities to treat a surface so that bonding occurs between it and other surfaces, e.g., wood and thermoplastics." The effects of coupling agents on the mechanical properties of WPCs have been studied extensively (Woodhams et al. 1984; Maldas and Kokta

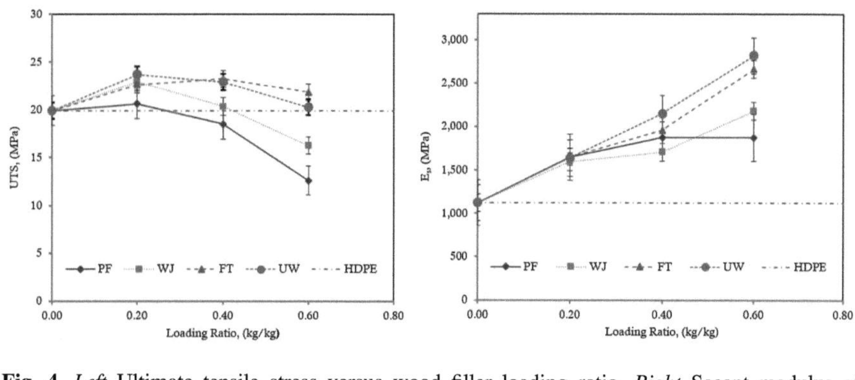

Fig. 4 *Left* Ultimate tensile stress versus wood filler loading ratio. *Right* Secant modulus of elasticity versus wood filler loading ratio; *PF* pine flour, *WJ* whole-tree juniper, *FT* forest thinning, *UW* urban wood, *HDPE* reference specimen with no filler (Karas 2010)

1991; Raj and Kokta 1991; Stark and Rowlands 2003) and have shown that coupling agents increase the strength and stiffness of the bulk composite. This approach has been the topic of much research and a detailed review of coupling agents used in WPCs has been compiled by Lu et al. (2000). One commonly used coupling agent is maleic anhydride-grafted polypropylene (MAPP). This type of coupling agent reacts with the wood component on one end, and on the other end entangles a modified PP with the bulk PP polymer. Stark and Rowlands (Stark and Rowlands 2003) study showed that the addition of MAPP had the greatest effect on the properties of wood fibre composites containing wood particles with greater aspect ratios ($\approx$16). As they pointed out, wood particles commonly used in WPCs have low aspect ratios (3–5). This being the case, coupling agents can only assist in interfacial bonding to a limited extent. While one would expect better performance from fibres with larger aspect ratios, this method adds cost and complexity to the manufacture and processing of WPCs. Whether or not a WPC manufacturer decides to use a coupling agent, the interaction between the polymer matrix and the embedded particle still requires attention. Unlike measuring mechanical properties of WPCs like creep or bending strength at a macroscale, understanding the interaction between the particle and matrix requires a look at the microscale. In the past, a variety of methods have been used to explain and predict the interactions between the wood and polymer phases using idealised analytical and numerical techniques (Clyne 1989). These methods held some common assumptions including: the embedded particles which are homogenous and isotropic, impermeable, cylindrical in shape, have a large aspect ratio, have a perfect bonding interface with the matrix, and have no transfer of load on their ends. Such assumptions can hardly be applied to irregular, porous bio-based particles like wood (Raisanen et al. 1997). These methods provided approximations of load transferred with an embedded inclusion in a thermoplastic matrix but lacked the complexity of the actual system to be satisfactory. Recently, Schwarzkopf and Muszynski (2015) investigated these interactions using optical measurement techniques based on the digital image correlation (DIC) principle. This study aimed to develop a methodology for the efficient measurement of strain distribution patterns in the matrix material surrounding embedded wood particles. Wood particles and reference wire particles were embedded in a HDPE matrix. The specimens were pulled in tension and imaged throughout the test. By comparing successive images to one another, the displacements and strains on the surface of the specimen could be determined. The results from their study showed that there is a good agreement between theoretical (Clyne 1989) and observed strain distribution patterns (Fig. 5). However, a quantitative analysis of the load transfer between the two needs to happen using morphologically informed predictive modelling tools. This approach has been used for analysing load transfer in adhesively bonded specimens by Kamke et al. (2014) and gives a unique look at the internal stress transfer in wood-based composites.

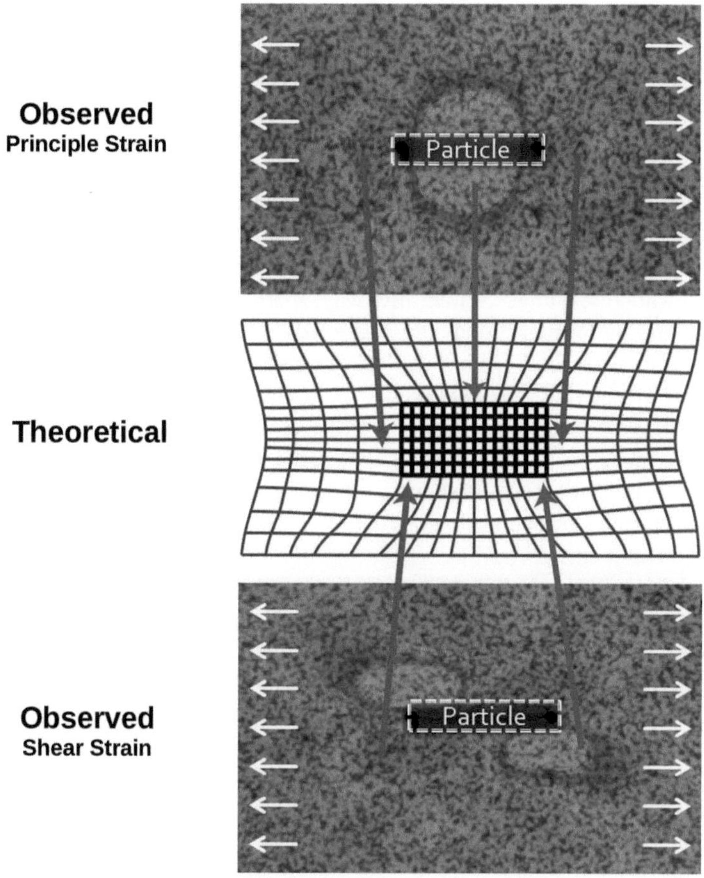

Fig. 5 Comparison of observed strains (obtained from DIC) and a theoretical representation of the interaction between an embedded particle and a polymer matrix. Theoretical model adapted from Clyne (1989)

4.2 Durability

When compared with solid wood materials, WPCs have lower mechanical properties in strength, stiffness, and creep resistance. On the other hand, WPCs are less susceptible to moisture absorption and absorb at a slower rate, providing for better resistance to fungal attack and dimensional changes (Caulfield et al. 2005). This being the case, it is important to remember that the wood particles within the WPC are still a nutrient source for microbial decay and have the potential to become degraded. While in theory the encapsulation of wood particles by a polymer should provide some level of protection from biological decay and moisture intake, external forces may compromise this protective layer.

WPCs are often marketed as highly durable, low maintenance, and a good alternative to solid wood as an exterior product. One area under investigation for using WPCs in exterior applications is in highway signs and markers. There are many types of roadway markers, signs, and fixtures that exist along every kilometre of the roadway.

Based on only two of these markers, tubular markers and inroad reflectors (Fig. 6), Thompson et al. (2010) estimated that approximately 870 t of plastic is being used annually in eight western states of the US. By introducing any amount of wood filler into these products, a substantial volume of plastic could be displaced by a renewable resource. While highway markers do not generally demand high-structural performance, they are subjected to harsh environmental conditions, mechanical abrasion, and ground contact throughout their entire service life. Rain will soak the materials, temperatures will drop below and well above freezing temperatures, vehicle tyres will drive over inroad markers, and the sun will expose them to UV rays. Karas (2010) investigated the use of WPCs in highway applications and assessed their durability. In this study, durability was assessed by measuring selected mechanical properties of WPCs before and after accelerated weathering treatments as well as weight loss due to ground contact tests. To simulate the effects of environmental conditions that WPCs used in highway applications encounter, an accelerated weathering unit was used that applied UV light, heat, and moisture. A soil contact test was also used to assess the durability of the WPCs with respect to mass loss from biological decay.

As expected, WPC specimens that had been exposed to accelerated weathering experienced lower mechanical performance than those that were unweathered. Specimens that had higher loading levels of wood filler experienced more mass loss and in specimens that were exposed to accelerated weathering, the mass loss was almost double that of unweathered specimens. This was believed to be due to the degradation of the polymer at the surface of the specimens. While the HDPE was not substantially degraded due to biological decay in the soil contact tests, it may have been degraded when exposed to UV in the weathering treatments. By degrading the surface matrix material and allowing cracks and pathways to form throughout the structure, more wood particles were exposed to biological attack. This breakdown of the surface polymer encapsulating the wood fillers can also occur during freeze thaw cycles and due to physical wear by car and truck tyres abrading the surface of roadway markers. While WPCs used in decking materials will not need to protect against vehicles driving over them, the same durability concerns exist.

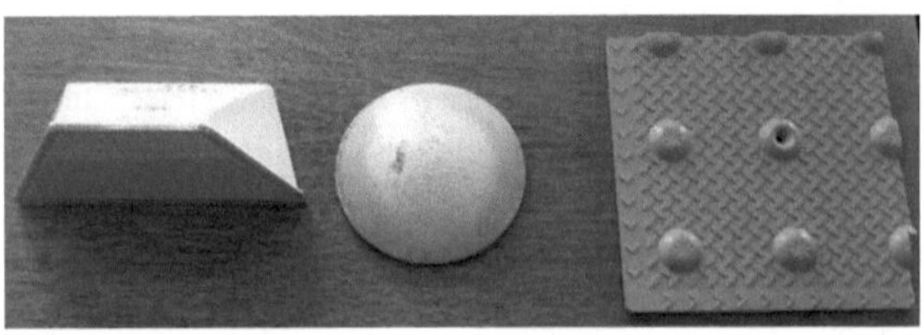

Fig. 6 Some examples of in-road markers found along the roadway. *Photo Credit* Karas, M

Schirp et al. (2008) provides a state of the art summary of the biological degradation of WPCs and provides some strategies for improving the durability of WPCs. One of these strategies is to limit the loading ratio of wood filler to under 50 % unless using an antimicrobial treatment such as zinc borate which is effective against wood-decay fungi and insects. This again touches on the incomplete encapsulation of wood particles at higher loading ratios. Essentially, any cracks or openings in the WPC or between the polymer and the wood provide a pathway for moisture and biological decay to enter, as well as for crack propagation. Another method used to improve the durability is to apply a cap stock layer. Most of the durability issues in WPCs start at the surface and work inwards. Instead of distributing expensive additives like antimicrobial, colourants, and anti-UV agents throughout the volume of the WPC, a thin layer of harder plastic with these protective additives is added to the outside (Hanawalt 2012). While effective, the cost of making the production process more complicated must be assessed.

5 WPC Applications and Use

In Europe, WPCs account for nearly 11 % (260,000 t) of composite products production, which includes product categories ranging from construction to consumer goods (Carus et al. 2015). Table 1 lists common products, their product categories, and the associated manufacturing processes typically used to create them. Outdoor decking and automotive components account for the greatest share of WPC production in Europe (67 and 24 %, respectively), while other uses including siding and fencing, furniture, consumer goods, and technical applications account for the remainder (Carus et al. 2015). WPCs are considered a growth market, with increases in major markets like North America and Europe estimated to be around 10 % while in China growth is estimated to reach 25 % in 2015 (Eder and Carus 2013).

Table 1 Some common WPC products, their product categories, and associated manufacturing products

Product examples	Product category	Manufacturing process
Decking boards and tiles, siding, and window frames	Construction, outdoors	Extrusion and injection moulding
Garden furniture and fencing	Garden/yard and outdoors	Extrusion
Automotive interior trims and engine components (exposed to temperatures less than 110 °C)	Automotive	Extrusion, injection moulding and sheet forming
Furniture parts and furniture	Housing and interior	Extrusion and injection moulding
Packaging (e.g. corner protectors), components for: games, household electronics and other devices	Consumer goods, interior and outdoor	Extrusion and injection moulding, FDM

Table adapted from Eder and Carus (2013)

5.1 *Construction*

WPCs are often used in the built environment, both indoors and outdoors. Although exterior decking is the most common application worldwide, other construction-related products are also important. These include railing system components, stairs, window and door applications, flooring, exterior siding, fencing and landscape materials, interior moulding and trim work (Clemons 2002; Klyosov 2007; Eder and Carus 2013; Carus et al. 2015). WPCs for construction purposes are typically extruded and made to function and look like their solid wood counterparts. These WPCs support conventional fasteners like screws, nails, brackets, and others which allow them to be used without significant adjustment to typical building practices (Pritchard 2004). The appearance of WPCs depend on the material contents and finishing measures which include imprinting "wood-like" properties such as grain structure and coatings, as in Fig. 7.

5.2 *Automotive*

The automotive industry uses WPC and other renewable-based fibre composites in increasing quantities both to offset costs, reduce weight, and, in Europe, in an effort to meet the demands for high recyclability contents of vehicles. Wood fibres in fibre–plastic composites used in the automotive industry account for only about 38 % of the total renewable fibre usage (Carus et al. 2015). Other fibres in common use are cotton, flax, kenaf and hemp. The primary use of WPC's in automobiles is for storage (trims in trunks, shelves for spare tyres, etc.) and interior door trims, while other renewable fibre-based composites are used for higher value interior

Fig. 7 WPC coloured and formed to imitate the appearance of wood

trims including in dashboards and doors (Carus et al. 2015). In contrast to other fields, WPCs in the automotive field are often used to replace metal or fibreglass components, as opposed to wood or plastics.

European regulations stipulate that by 2015 reuse and recovery of all end-of-life vehicles in the European Union must reach a minimum of 95 % by average vehicle weight, and reuse and recycling of materials from end-of-life vehicles must reach a minimum of 85 % by average vehicle weight (The European Parliament and the Council of the European Union 2000). The use of renewable fibre-based composites significantly reduces the weight of many vehicle components, and also increases the recyclability. These factors, along with the existing technological capabilities of vehicle manufacturers, provide for strong future growth in the automotive sector, particularly in Europe where renewable fibre use could increase from 60,000 to 300,000 t each year (Carus et al. 2015).

5.3 Furniture, Highway Materials, Consumer Goods, and Other WPC Applications

WPCs are already in use by major furniture producers, such as IKEA (2015a, b). Applications in packaging and consumer products, including instruments, toys, tableware are not yet mainstream, but are areas of likely growth (Haider and Eder 2010).

Highway construction in the United States utilises several materials which are suitable for substitution by WPCs (e.g. treated wood and pure plastics), and may become a growth sector in the future if certain barriers are overcome (Thompson et al. 2010).

5.4 Perceptions of WPCs

User perceptions, knowledge, and preferences regarding materials have a large impact on the overall utilisation of a product and vary in importance and impact based on the needs and expectations of the user (Clemons 2002; Thompson et al. 2010). Material specifiers, manufacturers, and product users (both builders and building occupants) all interact with materials and form opinions from their experiences, which ultimately impact later material specification decisions.

Manufacturers of products, especially of products that are exposed to public view like furniture, interior trim work, exterior decking, etc., may perceive WPCs as ‘fake’ and undesirable despite their workability and potential to imitate the appearance of wood (Pritchard 2004). If manufacturers do not approve of the material and refuse to use it (or do so begrudgingly), WPCs are unlikely to gain traction in this subsector of the wood products market. However, manufacturers who typically deal with plastics are more likely to view the material in positive terms because it reduces material costs compared to solid plastics and can be considered to improve appearances, and reduce environmental impacts (Pritchard 2004; Bismarck et al. 2006; Klyosov 2007).

In the highway construction sector in the western U.S., WPCs were perceived favourably by highway contractors compared to other more frequently used materials despite low utilisation and relatively low familiarity (Thompson et al. 2010). Given the favourable perception of these products, the authors conclude that once other challenges related to product certification are met, WPCs are well placed to enter a potentially lucrative and sizeable market (Thompson et al. 2010).

In contrast to manufacturers and builders, lay persons often have different needs and expectations from materials, especially related to aesthetics and maintainability. For example, day-to-day building users have both active and passive responses to their built environment. In both cases, the materials, design, and perceived qualities of the environment impact user's subjective and biological responses to the environment (Burnard and Kutnar 2015). In the case of subjective reactions users may decide that they dislike a building's (or products) design and in some cases may choose not to use it which has direct economic impacts. In the case of biological responses, human well-being can be impacted by the user's reaction to their environment, especially over long periods of time. Material selection is a critical part of designing buildings for human well-being (Burnard and Kutnar 2015). In new building design paradigms that emphasise providing positive human health impacts, including nature and natural elements in the built environment is important because positive health impacts have been demonstrated repeatedly in outdoor environments with greater degrees of naturalness (less apparent human intervention). User perceptions of material naturalness are then a key to determine which materials should be used in buildings designed to produce positive health impacts for occupants (Burnard et al. 2015). A study comparing user perceptions of building material naturalness conducted in three countries (Finland, Norway, and Slovenia) demonstrated that users were quickly able to identify WPCs with imitated wood-like features as significantly less natural than wood products that had undergone varying degrees of transformation (solid wood, wood-based composites, stone, metal, ceramics, and textiles were included in the study) (Burnard et al. 2015). However, the WPC in the study was one of the few materials where differences in perceptions of naturalness between countries were apparent, indicating that participants from two of the countries (Finland, Norway) were able to distinguish between the natural and imitated materials more readily than their counterparts (Slovenia). In addition to less favourable ratings of naturalness, the study found that users in Finland and Norway view the material as natural in a binary decision task about half as often as Slovenians (25 and 50 %, respectively) (Burnard et al. 2015). Compared to other materials in the study only seven materials were viewed as not natural by more respondents overall (Fig. 8). In these cases, the difference in perceptions may be related to the more prevalent use of wood materials in the built environment in Finland and Norway than in Slovenia (Burnard et al. 2015).

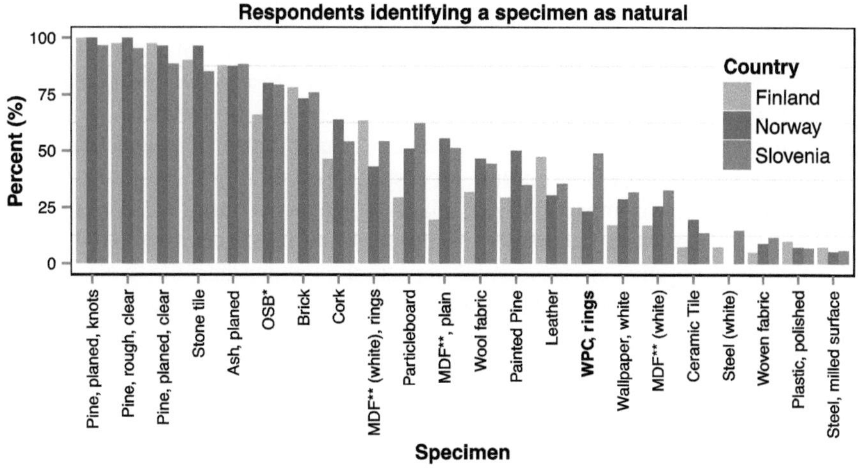

Fig. 8 Percent of respondents stating the listed specimen was natural in a binary decision assessment (Adapted from Burnard et al. 2015). **OSB* Oriented Strand Board; ***MDF* Medium Density Fibreboard

6 Environmental Impacts of WPCs

Environmental impacts of processes and products have become an increasingly publicised and are used by scientists, marketers, material selectors, and the public to gain insight into how different technologies and products affect our environment. Life cycle assessment (LCA) is means of estimating the environmental impact of a product and the process used to create it. LCAs examine inputs and outputs of a system used to create a product over a specified portion of its life. For example, a cradle-to-grave LCA would use detailed information about the production and extraction of raw materials, transportation, manufacturing, throughout the product's useful life, and finally disposal (with wood products final disposal includes significant energy recovery through combustion). Cradle-to-gate is another common assessment period, which begins with raw material production but ends when the product leaves the producers facility and does not include the use phase or disposal. However, with the increased focus on recycling and reuse accounting for the environmental impact of products from cradle-to-cradle is increasingly important, especially in Europe where recycling and environmental targets are quite aggressive (Gärtner et al. 2012; Höglmeier et al. 2013). In cradle-to-cradle scenarios for renewables, materials are "cascaded" and reused after each life cycle, preferably with minimal transformation in sequential uses, until finally the materials are burned for energy reclamation (Gärtner et al. 2012; Höglmeier et al. 2013). WPCs fit well into the wood cascade as repeat processing continually produces material side streams which are suitable for use in WPCs, and reduction in material quality and size (from solid timber to particles and fibres) after each life cycle will eventually produce materials suitable for use in WPCs. Recycled plastics are also used in WPCs which can significantly reduce their environmental impact (Vidal et al. 2009; Oneil et al. 2013). Not only does wood use offset the quantity of non-renewable material in WPCs, using wood in long-life products also sequesters atmospheric carbon and mitigates the climate change potential connected to the energy use of material extraction and processing (Hill et al. 2015).

The environmental impact of WPC matrix components varies greatly based on the type of matrix used. Commonly used petroleum-based polymers (HDPE, PP, PVC) produce negative environmental impacts throughout their life cycle, largely because they rely on non-renewable raw materials (Rajendran et al. 2012; Qiang et al. 2014). Using alternative, renewable-based, biodegradable polymers such as PLA may produce smaller environmental impacts once improved manufacturing techniques are developed that reduce water and energy utilisation (Qiang et al. 2014).

Studies directly comparing solid timber decking products against alternative WPCs conclude that solid wood produces a significantly lower environmental impacts than in both cradle-to-gate and cradle-to-grave scenarios (Bolin and Smith 2011; Oneil et al. 2013). Bolin and Smith (2011), following ISO 14040 and ISO 14044 standards (ISO/IEC 2006a, b), compared treated lumber used for decking (alkaline copper quaternary (ACQ) lumber, an unspecified southern pine from the US) to a WPC composed of 50 % recycled wood with a matrix mixture of virgin and recycled HDPE (25 % of the total each). Comparing a theoretical deck of approximately 30 m^2, the ACQ-treated lumber deck used 14 times less fossil fuels and nearly 3 times less water, produced 4 times less acid rain, 3 times less greenhouse gas emissions, half of the smog and ecological toxicity, and nearly equal eutrophication (Bolin and Smith 2011). The authors note that using biomass or renewable energy sources for WPC production could significantly reduce the striking difference in fossil fuel consumption between the ACQ-treated lumber and the WPC (Bolin and Smith 2011). Assuming production in regions where renewable energy sources are more common (Europe, or the west coast of the US) could significantly alter the outcomes of this type of comparison.

Another study comparing 9.3 m^2 of decking made from virgin PVC, virgin WPC, recycled WPC, and redwood lumber found that recycled virgin WPC produced less global warming potential and ozone depletion than PVC, more smog, acidification, eutrophication and respiratory effects using the US Environmental Protection Agency's TRACI methodology (Oneil et al. 2013). Recycled WPC produced reduced impacts in comparison to virgin WPC in all categories except ozone depletion, in which it produced similar effects. Compared to PVC, recycled WPC produced significantly reduced impacts except eutrophication which was greater for recycled WPC and smog impacts, which were approximately equal for both (Oneil et al. 2013). In all categories, redwood lumber produced significantly lower impacts between 10 and 30 % of the most significant impacts of the compared products, except for global warming potential that produced a negative effect (approximately 140 % reduced impact compared to PVC) (Oneil et al. 2013).

6.1 Product Category Rules and Environmental Product Declarations

Product category rules (PCR) are defined methods for creating environmental product declarations (EPD) for product groups such as "wood materials". EPDs are standardised statements of products environmental impacts that are meant to be comparable to other similarly produced products that have EPDs following the same PCR. The international standard for producing PCRs is ISO 14025, which defines the necessary elements of PCRs, and, in effect, of EPDs (ISO/IEC 2006c). Though no specific PCR or EPD could be found for a WPC product of any kind at the time of this writing, a PCR for general wood materials can be used to produce a WPC EPD. The PCR for wood materials developed by the firm Institut Bauen und Umwelt e.V., specifies that their PCR can be used for a variety of wood composites, including special wood materials which could include WPCs (Institut Bauen und Umwelt e.V. 2009). Going forward using actual products for comparison, using LCAs or EPDs-produced following a published PCR will provide more comparable results. However, these tools are still developing and authors of LCAs and EPDs should utilise the latest standards and versions of PCRs to conduct their comparisons.

7 Summary

WPCs are a product category with many existing and emerging applications. The vast majority of WPCs are used for exterior decking and other exterior board applications. This market continues to grow in North America, is beginning to grow in Europe, and rapid growth is expected in China. Automotive applications continue to grow as well, however, competition in this subsector from other fibre sources poses a risk for WPCs. In both these primary uses, wood is used to reduce the cost and weight of plastic products as well as to improve material properties such as strength.

An added benefit of using wood as a filler product in plastics is the reduced environmental impact that can be achieved by offsetting the amount of non-renewable materials used in a product with renewably sourced wood. Often, the wood used in WPCs can come from primary production side streams, forest slash, or recovered wood products, which reduces the strain on raw forest resources. However, competition for these products from energy producers, as well as fibreboard and particleboard producers, has an impact on material costs for WPC manufacturers which may steer them towards other renewable sources instead of wood.

Current development is focused on improving material qualities ranging from strength to durability, as well as improving the compatibility of matrix components for longer lasting products. Other current work is focused on examining the environmental and performance impacts of using WPCs as substitution products in various new applications, using renewable-based plastics with wood reinforcement, and examining new manufacturing methods for WPCs.

Future research and innovation must overcome challenges related to matrix compatibility, rising costs related to increased demand for forest resources as a fuel source, product shortcomings such as exterior durability, impact strength, as well as knowledge gaps and negative perceptions amongst manufacturers in some sectors and consumers. Other areas for innovation and development are rapid manufacturing with WPCs either for prototyping or bespoke product manufacturing. In each case new innovations and developments must consider the environmental impacts associated with the raw materials, manufacturing process, product utilisation, recycling, and end-of-life scenarios. Improved data collection and quality amongst manufacturers and consumers will help to improve the quality of LCAs conducted on WPCs. Furthermore, using PCRs to produce comparable EPDs may alleviate negative consumer and manufacturers perceptions about the suitability and environmental impact of WPCs in a variety of product categories.

Acknowledgments The authors are pleased to acknowledge the support of WoodWisdom-Net+ and the Slovenian Ministry of Education, Science, and Sport of the Republic of Slovenia for their support of the What We Would believe and cascading recovered wood projects; European Commission for funding the project InnoRenew CoE under the Horizon2020 Widespread-2015 program, and infrastructure program IP-0035.

References

Berger MJ, Stark NM (1997) Investigations of species effects in an injection-molding-grade, wood-filled polypropylene. In: Fourth international conference on woodfiber–plastic composites, Madison, WI, pp 19–25

Bismarck A, Baltazar-Y-Jimenez A, Sarikakis K (2006) Green composites as panacea? Socio-economic aspects of green materials. Environ Dev 8:445–463. doi:10.1007/s10668-005-8506-5

Bodig J, Jayne B (1982) Mechanics of wood and wood composites. Van Nostrand Reinhold, New York

Bolin C, Smith S (2011) Life cycle assessment of ACQ-treated lumber with comparison to wood plastic composite decking. J Clean Prod 19:620–629. doi:10.1016/j.jclepro.2010.12.004

Bouafif H, Koubaa A, Perré P, Cloutier A (2009) Effects of fiber characteristics on the physical and mechanical properties of wood plastic composites. Compos Part A Appl Sci Manuf 40:1975–1981. doi:10.1016/j.compositesa.2009.06.003

Burnard MD, Kutnar A (2015) Wood and human stress in the built indoor environment: a review. Wood Sci Technol 49:969–986. doi:10.1007/s00226-015-0747-3

Burnard MD, Nyrud AQ, Bysheim K, Kutnar A, Vahtikari K, Hughes M (2015) Building material naturalness: perceptions from Finland, Norway and Slovenia. Indoor Built Environ 0:1–16. doi:10.1177/1420326X15605162

Carus M, Eder A, Dammer L, Korte H, Scholz L, Essel R, Breitmayer E, Barth M (2015) Wood-plastic composites (WPC) and natural fibre composites (NFC): European and global markets 2012 and future trends in automotive and construction

Caulfield DF, Clemons C, Jacobson RE, Rowell RM (2005) Wood thermoplastic composites. In: Rowell RM (ed) Handbook of wood chemistry and wood composites. CRC Press, Boca Raton, pp 365–378

Clemons C (2002) Wood-plastic composites in the United States: the interfacing of two industries. For Prod J 52:10–18

Clyne TW (1989) A simple development of the shear lag theory appropriate for composites with a relatively small modulus mismatch. Mater Sci Eng A122:183–192

Corbière-Nicollier T, Gfeller Laban B, Lundquist L, Leterrier Y, Månson J-A, Jolliet O (2001) Life cycle assessment of biofibres replacing glass fibres as reinforcement in plastics. Resour Conserv Recycl 33:267–287. doi:10.1016/S0921-3449(01)00089-1

Eckert C (2000) Opportunities for natural fibers in plastic composites. In: Progress in woodfibre–plastic composites. University of Toronto

Eder A, Carus M (2013) Global trends in composites (WPC). Bioplastics Mag 8:16–17

English B, Stark N, Clemons C (1996) Wood and mineral fillers for injection molding grade polypropylene, Madison, WI

Escobar WG, Wolcott MP (2008) Influence of wood species on properties of wood/HDPE composites. Washington State University

Farsi M (2012) Thermoplastic matrix reinforced with natural fibers: a study on interfacial behavior. In: Wang J (ed) Some critical issues for injection molding. InTech, pp 225–250

Gacitua W, Wolcott MP (2009) Morphology of wood species affecting wood-thermoplastic interaction: microstucture and mechanical adhesion. Maderas-Ciencia y Technol 11:217–231

Gärtner SO, Hienz G, Keller H, Paulsch D (2012) Ökobilanz der kaskadierten Nutzung nachwachsender Rohstoffe am Beispiel Holz– eine Einordnung = "LCA of cascading use of renewable resources on the example of wood - a classification". uwf. UmweltWirtschaftsForum 20:155–164. doi:10.1007/s00550-012-0259-7

Guo Y, Zeng W, Jiang K (2011) Preparation and selective laser sintering of wood–plastic composite powers and post processing. Dig J Nanomater Biostruct 6:1435–1444

Haider A, Eder A (2010) Markets, applications, and processes for wood polymer composites (WPC) in Europe. In: Teischinger A, Barbu MC, Dunky M, Harper D, Jungmeier G, Militz H, Musso M, Petutschnigg A, Pizzi A, Wieland S, Young TM (eds) Processing technologies for the forest and biobased product industries, pp 146–151

Hanawalt K (2012) Wood-pastics composites done right. Plast Technol 32–34

Hill C, Norton A, Kutnar A (2015) Environmental impacts of wood composites and legislative obligations. In: Wood composites, pp 309–332

Höglmeier K, Weber-Blaschke G, Richter K (2013) Potentials for cascading of recovered wood from building deconstruction—a case study for south-east Germany. Resour Conserv Recycl 78:81–91. doi:10.1016/j.resconrec.2013.07.004

Hull D, Clyne TW (1996) An introduction to composite materials, 2nd edn. Cambridge University Press, Cambridge

IKEA (2015a) Democratic design—IKEA

IKEA (2015b) IKEA PS 2012 Chair with armrests—IKEA

Institut Bauen und Umwelt e.V. (2009) PCR Wood Materials. Königswinter

ISO/IEC (2006a) ISO/IEC 14040:2006 environmental management—life cycle assessment—principles and framework

ISO/IEC (2006b) ISO/IEC 14044:2006 environmental management—life cycle assessment—requirements and guidelines

ISO/IEC (2006c) ISO/IEC 14025:2006 environmental labels and declarations—type III environmental declarations—principles and procedures

Kamke FA, Nairn JA, Muszynski L, Paris JL, Schwarzkopf M, Xiao X (2014) Methodology for micromechanical analysis of wood adhesive bonds using x-ray computed tomography and numerical modeling. Wood Fiber Sci 46:15–28

Karas M (2010) Sustainable bio-composites for west coast highways. MS thesis, Oregon State University

Klyosov A (2007) Wood-Plastic composites. Wiley, Hoboken

La Mantia F, Morreale M (2011) Green composites: a brief review. Compos Part A Appl Sci Manuf 42:579–588. doi:10.1016/j.compositesa.2011.01.017

Lu JZ, Wu Q, McNabb HS Jr (2000) Chemical coupling in wood fiber and polymer composites: a review of coupling agents and treatments. Wood Fiber Sci 32:88–104

Maldas D, Kokta BV (1991) Influence of maleic anhydride as a coupling agent on the performance of wood fiber—polystyrene composites. Polym Eng Sci 31:1351–1357

Mukherjee T, Kao N (2011) PLA based biopolymer reinforced with natural fibre: a review. J Polym Environ 19:714–725. doi:10.1007/s10924-011-0320-6

Nikzad M, Masood SH, Sbarski I (2011) Thermo-mechanical properties of a highly filled polymeric composites for fused deposition modeling. Mater Des 32:3448–3456. doi:10.1016/j.matdes.2011.01.056

Oneil E, Bergman R, Han H, Eastin I (2013) Comparative life-cycle assessment of California redwood decking

Patterson J (2001) New opportunities with wood-flour-foamed PVC. J Vinyl Addit Technol 7:138–141. doi:10.1002/vnl.10281

Pritchard G (1998) Plastics additives: an A–Z reference. Springer Science and Business Media, Berlin

Pritchard G (2004) Two technologies merge: wood plastic composites Geoff Pritchard describes how wood and resin are being. Plast Addit Compd 48:18–21. doi:10.1016/S0034-3617(04)00339-X

Qiang T, Yu D, Zhang A, Gao H, Li Z, Liu Z, Chen W, Han Z (2014) Life cycle assessment on polylactide-based wood plastic composites toughened with polyhydroxyalkanoates. J Clean Prod 66:139–145. doi:10.1016/j.jclepro.2013.11.074

Raisanen VI, Alava MJ, Niskanen KJ, Nieminen RM (1997) Does the shear-lag model apply to random fiber networks? J Mater Res 12:2725–2732

Raj RG, Kokta BV (1991) Reinforcing high density polyethylene with cellulosic fibers. I: The effect of additives on fiber dispersion and mechanical properties. Polym Eng Sci 31:1358–1362

Rajendran S, Hodzic A, Soutis C, MariamAl-Maadeed A (2012) Review of life cycle assessment on polyolefins and related materials. Plast Rubber Compos 41:159–168. doi:10.1179/1743289811Y.0000000051

Schirp A, Stender J (2009) Properties of extruded wood-plastic composites based on refiner wood fibres (TMP fibres) and hemp fibres. Eur J Wood Prod 68:219–231. doi:10.1007/s00107-009-0372-7

Schirp A, Ibach RE, Pendleton DE, Wolcott MP (2008) Biological degradation of wood-plastic composites (WPC) and strategies for improving the resistance of WPC against biological decay. ACS Symp Ser 982:480–507. doi:10.1021/bk-2008-0982.ch029

Schwarzkopf MJ (2014) Characterization of load transfer in wood-based composites. PhD Dissertation, Oregon State University

Schwarzkopf M, Muszyński L (2015) Strain distribution and load transfer in the polymer-wood particle bond in wood plastic composites. Holzforschung 69:53–60. doi:10.1515/hf-2013-0243

Sjöström E (1993) Wood chemistry: fundamentals and applications. Academic Press, San Diego

Stark NM, Rowlands RE (2003) Effects of wood fiber characteristics on mechanical properties of wood/polypropylene composites. Wood Fiber Sci 35:167–174

Teuber L, Militz H, Krause A (2015) Processing of wood plastic composites: the influence of feeding method and polymer melt flow rate on particle degradation. J Appl Polym Sci 43231. doi:10.1002/app.43231

The European Parliament and the Council of the European Union (2000) Directive 2000/53/EC—end-of-life vehicles

Thompson DW, Hansen EN, Knowles C, Muszynski L (2010) Opportunities for wood plastic composite products in the U.S. highway construction sector. Bio Resour 5:1336–1352

Vidal R, Martínez P, Garraín D (2009) Life cycle assessment of composite materials made of recycled thermoplastics combined with rice husks and cotton linters. Int J Life Cycle Assess 14:73–82. doi:10.1007/s11367-008-0043-7

Wang Y (2007) Morphological characterization of wood plastic composite (WPC) with advanced imaging tools: developing methodologies for reliable phase and internal damage characterization. Oregon State University

Westman MP, Fifield LS, Simmons KL, Laddha SG, Kafentzis TA (2010) Natural fiber composites: a review. Pacific Northwest National Laboratory, Report PNNL—19220
Wolcott MP, Englund K (1999) A technology review of wood-plastic composites. In: 33rd international particleboard materials symposium, pp 103–111
Woodhams RT, Thomas G, Rodgers DK (1984) Wood fibers as reinforcing fillers for polyolefins. Polym Eng Sci 24:1166–1171. doi:10.1002/pen.760241504
Zhong W, Li F, Zhang Z, Song L, Li Z (2001) Short fiber reinforced composites for fused deposition modeling. Mater Sci Eng A 301:125–130. doi:10.1016/S0921-5093(00)01810-4

The Opportunities and Challenges for Re-use and Recycling of Timber and Wood Products Within the Construction Sector

Graham A. Ormondroyd, Morwenna J. Spear and Campbell Skinner

Abstract This chapter addresses the many factors which have influenced the re-use and recycling of wood in the UK over approximately a 20 year period to 2015. Drawing on a wide range of reports and data for the sector, the main trends are described. These include early adoption of standards for the quality of recycled wood in the particleboard industry, clear segregation of waste wood by origin and level of expected contamination, and more recently, the development of the biomass energy market. The construction sector presents several challenges when reclaiming, re-using or recycling timber, however, significant progress has been made by the introduced legislation, and through initiatives, best practice and the development of waste transfer stations and businesses utilising waste wood. Further avenues of research and emerging technologies are also discussed.

Keywords Construction · Demolition · Wood · Recycling · Reuse · Cascading use · Life cycle assessment (LCA) · United kingdom

1 Introduction

Construction and demolition (C&D) waste accounts for almost 30 % of all waste generated within the EU, and is a high priority for waste reduction measures, such as recycling and reuse. This waste stream consists of a mixture of stone, bricks, gypsum, cement, plastics, solvents, asbestos and wood, however, the dominant portion is the inert masonry and concrete. When considered on a volume basis timber occupies a slightly greater portion, so cannot be overlooked, and it is timber from C&D activities which is the focus for this chapter.

G.A. Ormondroyd (✉) · M.J. Spear · C. Skinner
The BioComposites Centre, Bangor University, Deiniol Rd, Bangor LL57 2UW, UK
e-mail: g.ormondroyd@bangor.ac.uk

G.A. Ormondroyd
Department of Architecture and Civil Engineering, University of Bath, Claverton Down, Bath BA2 7AY, UK

A. Kutnar and S.S. Muthu (eds.), *Environmental Impacts of Traditional and Innovative Forest-based Bioproducts*, Environmental Footprints and Eco-design of Products and Processes, DOI 10.1007/978-981-10-0655-5_3

Flows of recycled and reclaimed materials are complex, and become further complicated if regional influences are not minimised. As a result the data and examples used here relate largely to the UK context, with only small reference to best practice where it occurs or has been modelled in other regions. Examples have been drawn from various important points within the flow of wood waste through different industries with the aim of highlighting important blocks or restrictions to flow from one sector to another; but also to demonstrate the good practice and progress which has been made since the intention to reduce waste to landfill was highlighted as a policy concern within the UK, and within Europe as a whole.

The chapter, by using the UK as an extended example, sets out to develop an understanding of the use of timber in construction and the potential for the reuse and recycling of timber from this sector. It then goes on to discuss the current recycling and reuse pathways available for demolition wood and concludes by looking to the future and possible routes to increase the use of demolition wood within and beyond the construction industry.

2 Timber in Construction in the UK

2.1 Historical Use

The use of timber in construction in the UK has undergone several phases, which are reflected in the broad range of buildings and architecture from different eras still in service today. Mediaeval buildings relied on hardwood timbers for structural elements, with wattle and daub infill (Fig. 1). Other materials were used in different regions, depending on supply such as cob walls (earth and straw), rubble, flint, slate or chalk. As these buildings are generally few in number and frequently listed for protection or conservation, they contribute little to the recycled wood sector.

There was a shift in construction materials during the 1600s, with an increasing masonry component due to its lower risk in fire, and a change in timber availability and cost. Brick and stone had become common by the Georgian period (1715–1830) and continued to be dominant through to Victorian and Edwardian Britain (Fig. 2). Brick was also used in significant volumes for the many mills, works, warehouses and infrastructure of the industrial revolution, alongside the distance spanning capability of cast iron posts and arches. Later steel offered a further shift in performance and engineering, with girders, trusses and applications requiring tensile forces (Fig. 3). The majority of these buildings still relied on timber for its distance spanning properties in floors and roofs so still contribute a significant volume of timber if demolished. The timber forms the interior structures (joists, floor boards) and the roofing timbers (ridge board, purlins, rafters and sarking boards where utilised). These members are predominantly softwood, and typically may be pitch pine and deal from first growth forests, with relatively high strength, and well seasoned.

Fig. 1 Early buildings in Britain frequently used wood as the primary structural elements

Twentieth century construction continued to rely on brick and block for domestic buildings, and included the traditional timber roofing and flooring components (Fig. 2). A later shift to trussed rafters altered the dimensions and strength properties required for timber, but maintained the presence of timber in domestic buildings. At the same time, other materials were explored. Concrete was used for various projects, such as pre-fabricated housing, or public buildings. Industrial buildings dating from the twentieth century have utilised a range of building materials and styles, initially with a greater proportion of glazing and concrete than the Victorian brick structures, but later a shift towards sheet metal on a steel frame, or cement structures for heavy duty applications. In some of these constructions there is a negligible timber component (Fig. 4).

This broad mix of buildings and building materials contributes a diverse mixture of materials from demolition sites or re-development activities. Timber is present in the joists and flooring of the majority of buildings, and the roofing timbers of a

Fig. 2 Traditional brick and block construction contains a surprising volume of timber

significant number. In addition the joinery, fixtures and fittings may utilise a more diverse range of hardwoods and softwoods as well as more modern panel products where fitted furniture has been added or updated generation by generation. However, a significant proportion of waste from demolition sites remains aggregate (such as brick, stone and concrete), estimated at 30.8 million tonnes (Mt) in 2007 (based on 26.5 Mt reported by the NFDC members per year, 5.3 Mt by non-members, with 1Mt hazardous waste but the remainder being aggregate and inert as reported in CRWP Demolition Report (2009)).

Interestingly, audit data gathered by BRE for five demolition sites showed that the inert and aggregate materials tended to comprise up to 94 % of the building components by weight (Leisure centre, Table 1) with the average for all five audited buildings being 86.81 %). The timber component ranged from 1.45 % in the school buildings to 2.09 % by weight in the housing and flats (CRWP 2009). The timbers gathered from a demolition project will include a proportion of painted joinery

Fig. 3 The industrial revolution brought iron, then steel, into British construction methods

materials, some proportion of wood-based panels from interior fittings, as well as structural timbers of varying strength class. The strength of any timber present will depend to some extent on building age—Victorian buildings may contain pitch pine with typical strength and stiffness values approx. 50 % higher than redwood or domestically grown pine, while housing stock from the latter half of the twentieth century will have utilised imported redwood or whitewood for joists.

In addition to the timber from demolished or remodelled buildings, which accounts for approximately 25 % of the total wood waste in the UK (WRAP 2009), construction also produces approx. 26 % of the wood waste, due to off cuts necessary on site, over-ordered materials, or wastage due to sawing, planing and finishing over-sized materials or other finishing applications. A considerable quantity of timber is also used on the construction site for short- or medium-term applications, such as hoardings, temporary coverings and false work for concrete or cement. The construction activity itself therefore generates a significant proportion of wood waste (Figs. 5 and 6). A large proportion of this will be wood-based panels

Fig. 4 Modern buildings may not appear to contain much timber

(plywood or OSB), although treated timber may also be present in temporary site fencing, as will some softwood joinery timber, hardwood timber, particleboard and MDF offcuts depending on the style of interior fittings.

2.2 *Modern Methods of Construction*

During the final decades of the twentieth century, several factors drove the construction industry in the UK to consider newer methods and materials for domestic buildings, public/commercial buildings and other structures. One of the results was a shift towards timber frame construction, another was an increase in off-site construction methods, which may also include significant quantities of timber in components, such as structural insulated panels (SIPs), and pre-fabricated insulated

Table 1 Pre-demolition audits conducted by BRE for five types of building

Waste category	Building type			
	Housing and flats[a] (%)	Leisure (%)	School[b] (%)	Total (%)
Ceramics	12.08	12.17	1.88	6.55
Concrete	47.12	50.57	68.95	59.28
Inert	30.32	32.22	12.82	20.98
Insulation	0.00	0.00	0.08	0.05
Metals	7.76	2.99	12.62	9.98
Plastics	0.01	0.00	0.00	0.00
Timber	2.09	2.04	1.45	1.74
Plasterboard	0.63	0.00	2.19	1.42
Grand Total	100.0	100.0	100.0	100.0

[a]Two sites—social housing and private residences
[b]Data from two school buildings (CRWP 2009)

panels to infill in timber frame structures. Volumetric systems may also utilise timber, or other materials such as steel for frames of the pods assembled in the factory for transport to site. This means that future demolition waste profiles may include a greater proportion of timber, and of insulation products, dry lining boards and related materials with a decrease in brick, rubble, cement and the like.

In addition to the timber frame housing sector, the evolution of glued laminated timber in the 1950s and 1960s has resulted in a steady increase in the number of public buildings and other large developments (swimming pools, shopping centres, service stations, Fig. 7) using engineered wood. More recently interest in crossed laminated timber has developed, as this product is well suited to off-site construction methods, with rapid installation of pre-cut panels, and highly predictable strength and insulation properties. Even large buildings with steel or concrete structural supports may contain timber as cladding (Fig. 8).

Also significant within the past decade is the development of BREEAM assessments for commercial buildings, and the Code for Sustainable Homes for housing developments. Other schemes such as LEED (Leadership in Energy and Environmental Design) have also had a role in altering the levels of performance and green credentials expected for materials in construction. The energy performance targets for buildings have also been steadily increased—resulting in greater levels of insulation and thermal efficiency within construction.

Ease of attaining thermal energy efficiency, and rapid construction are two reasons why timber frame construction has steadily increased as a proportion of new housing starts in the UK. In addition, timber is a carbon neutral raw material and has a low embodied energy (Davies 2009). Timber frame units reached 25.9 % of total UK housing starts in 2009 (Jeffree 2010), although a large difference exists between Scotland, where the proportion of timber frame units is nearer 75 %, and England and Wales, where estimates range from 6 to 17 % depending on publication source and year (Optimat 2002; UKTFA 2009).

Fig. 5 Wood products used on a renovation project include many temporary structures

The impetus to embrace off-site manufacture for house building is driven by the rapid construction once on-site using panels which have been formed off-site; the high energy efficiency which is possible in modular construction systems due to low thermal conductivity of the timber and infill for no extra increase in wall thickness; the skills shortages in traditional building methods; and lower waste and higher efficiency available when producing the panels and structural elements in a controlled environment. In addition, there has been increased effort to consider the circular economy, and it is proposed that modular buildings will allow for future disassembly, movement to a new site and re-erection within a new construction with lower levels of waste generated. Such designs are still in early stages, however, successes include modular and volumetric units prepared for temporary accommodation at large building projects, or for athletes/staff and security accommodation at sporting events such as the London 2012 Olympic Games.

Fig. 6 Wooden false work was used to cast concrete in this 1960s building

In 2007, the UKTFA reported that 54,400 homes and units were constructed using timber frame construction (Anon 2008). This represented around 0.33 million m^3 of timber (using the estimated value of 6 m^3 timber per unit as used in Davies (2009)). This figure represents only 3.1 % of the UK's softwood timber consumption at the time, but represents a significant quantity of sequestered carbon (estimated as 240 thousand tonnes), stored in service in the housing stock of the UK that year alone. The Read report estimates that the built environment actually stores 19 million tonnes of carbon (Read et al. 2009). As construction techniques such as timber frame, and SIPs increase their use within the market, this reserve of carbon stored in the timber components will increase. It also makes good sense to consider standardising labelling or post-occupancy information to allow a greater degree of re-use at the end of life (EOL). Construction techniques which support the circular economy are emerging.

The volume of timber present in traditional brick and block housing should not be overlooked (Fig. 9), as these buildings tend to use timber rafters and purlins, or more recently timber trusses as the structural elements of the roof. Timber joists will be present throughout, and OSB or plywood sheathing and flooring. In total, the construction sector consumed 4.9 million m^3 of timber in 2009, however, the bulk

Fig. 7 Some modern buildings may contain large dimension timber structural elements, such as this glulam

of this was used in repairs, maintenance and improvements (RMI, 4.577 million m^3) and only 311 thousand m^3 in new build (Moore 2010). The RMI sector does include a significant structural component, in the form of home extensions with flat or pitched roofs, replacement of interior studwork walls, etc. Softwood used for fencing and outdoor timbers are reported separately, and comprised 1.056 million m^3 in 2009 (Moore 2010).

The total UK consumption of hardwoods is much lower than softwoods, and as a result, the total volume of hardwood in a typical new build house is negligible. Selected tropical or temperate hardwoods may be used for interior joinery or window frames. Depending on fashions at the time various hardwoods may be used in fitted kitchens or fitted furniture. On the other hand, there has been a steady uptake of solid frame oak construction, using traditional framing systems based on half-timbered barn designs from the fifteenth to seventeenth century. This is often given a modern level of regulation, structural calculations and energy efficiency measures, but utilises large section beams and pegged mortise and tenon joints. The modern oak-framed building is well insulated and may be clad in any suitable material, leaving only certain pre-decided timber members exposed for their aesthetic benefits (the remaining structural elements being within the make-up of the wall). Elements may be 10 cm or more in thickness or width, allowing the necessary carpenters joints to be formed. Oak is the favoured hardwood timber for such

Fig. 8 Other buildings may use timber as a cladding rather than a structural component

buildings, but similar designs using sweet chestnut have been demonstrated in some regions. These bespoke buildings are unlikely to contribute immediately to the recycling of timber from construction, however future generations are likely to be able to sell reclaimed beams from them in the same manner as the beams of large farmhouses, chapels, barn conversions are traded today. This is predominantly through private sale, either in person or online, but may also pass through the hands of a trader or merchant with suitable space for stock holding.

In the commercial and public buildings sector, but also in some housing developments, engineered wood in various forms has been growing in uptake. These include glulam for long span or large cross section beams and roofing elements; laminated veneer lumber (LVL) for high strength shorter span elements, and cross laminated timber (CLT). Relatively few such buildings are yet sufficiently old to require demolition, however, future generations may have additional options to consider in re-using or recycling these elements, as strength and uniformity are

Fig. 9 Typical brick and block construction awaiting demolition or remodelling

high, and large sections or wide panels will provide opportunities to re-saw to form new elements for the next phase of building.

Glulam has been available and in construction applications since the 1960s, so is likely to be the earliest to reach the attention of salvage merchants. It may be possible for these beams to be traded in the same manner as the large section timber beams from old Victorian buildings, either by private sale during remodelling activities, or via salvage merchants and online catalogues of available materials.

2.3 UK Waste Wood Market Size

The UK economy relies on wood for a wide range of products, supporting not only the construction and wood manufacturing sectors, but also civil engineering and composites manufacture where wood and wood-based panels are used to define moulds, forms and false work. The UK consumed 12.5 million tonnes of wood in 2007. This timber becomes waste during manufacturing operations, during short-term uses, or at end of life (EOL).

Data on waste wood generation within the UK has traditionally be poorly reported, as is indicated in the range of estimated values seen in Table 2 for studies conducted over a similar period by a variety of different methods. Effort to consolidate this data into a robust estimate was made by WRAP (2009), who compared values estimated from the top down and the bottom up, with relatively good correlation between the results. The WRAP (2009) data indicated that waste wood in the UK stood at 4.57 million tonnes per annum in 2007. This was reported between five sectors—construction; demolition and remodelling; packaging; industrial waste wood and municipal (from wood segregation facilities). The flow of wood consumed into wood waste and wood products is shown in the schematic (Fig. 10).

The typical breakout of these wood waste streams is indicated in Fig. 11. The bulk of this wood waste was generated or collected in England (3.906 million tonnes, 85 %), with Scotland generating 7 %, Wales 5 % and Northern Ireland 3 %.

Proportions of municipal wood varied between regions, depending on the level of provision for waste wood segregation at recycling centres and landfill sites, for example revealing strong regional trends. This sector is therefore likely to have shown the greatest degree of change over the following 8 years as greater provision for collection and segregation by local authorities has been implemented. For the 2007 data, a typical breakdown by region for construction wood waste was 84 % in England, 9 % in Scotland, 4 % in Wales and 3 % in Northern Ireland, whereas for municipal waste the split was 82 % in England, 9 % in Wales, 5 % in Scotland and 5 % in Northern Ireland (WRAP 2009). Within England, the North West, the South West and the North East had greatest provision for recycling municipal wood waste, with lowest provision in London. Municipal wood waste gathered from Wales and from the high performing regions of England is notably located close to wood-based panel manufacturers, providing an easy route to the next generation of wood product.

Table 2 Estimated values for timber waste within the different sectors, from four studies, in million tonnes (data and all reference taken from WRAP 2009)

Stream	ERM (2006), DTI (2007)	WRAP & MEL (2005)	TRADA (2002)	BRE (2004), Hurley (2004)
Municipal	1.1	1.1	2.5	0.8
Industrial and commercial	3.5	4.5	1.8	3.3
Construction/demolition and remodelling	2.9	5.0	0.9	3.3
Total	7.5	10.6	5.2	7.4

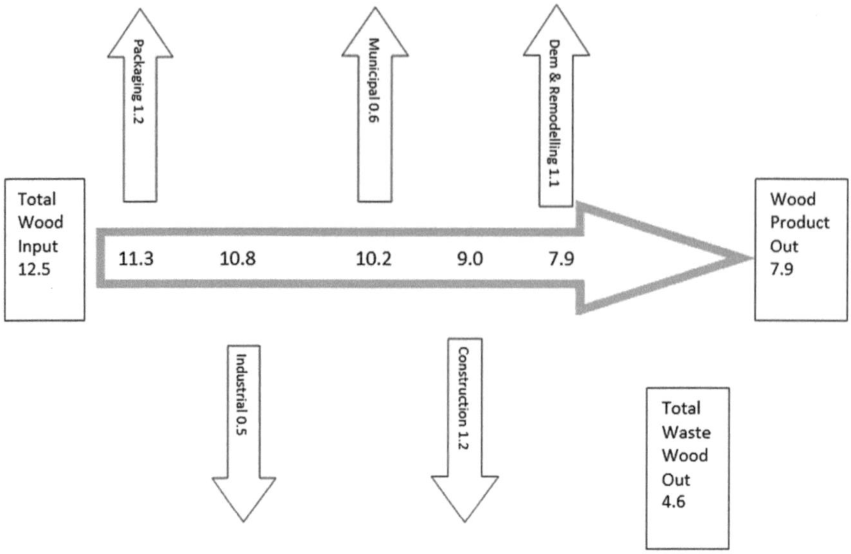

Fig. 10 Flow of waste generated by the five main wood using sectors in the UK in 2007, in million tonnes. *Data source* WRAP (2009)

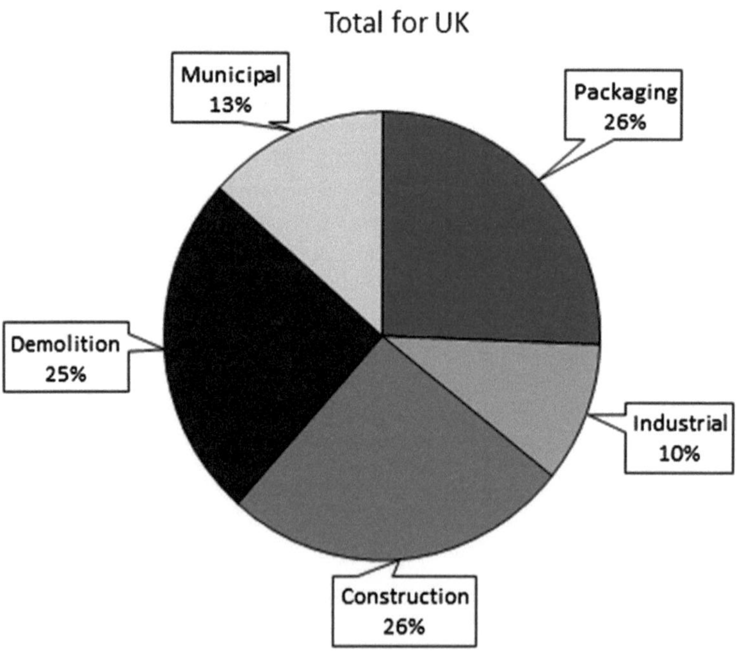

Fig. 11 Proportion of waste wood by sector within the UK in 2007 (WRAP 2009)

Table 3 WRA data and estimates for non-member organisations for tonnes of recycled wood entering different markets in 2007 (WRA 2008)

End market	WPA members	Others	Total
Panel board	817,000	383,000	1,200,000
Animal/poultry bedding	325,000	65,000	390,000
Equine surfaces	45,000	11,000	56,000
Mulches, soil conditioners, composts	60,000	15,000	75,000
Pathways and coverings	12,000	3,000	15,000
Biomass/energy	200,000	50,000	250,000
Total	1,459,000	527,000	1,986,000

Since this study, National Statistics have provided a breakout of wood waste in an annually produced digest of waste and resource statistics (DEFRA 2015). However, this data is split between the sectors where it is recovered, handled and produced. For example wood is 3.8 % of local authority collected waste (England 2010–11), which is 874 thousand tonnes; while wood is 19 % of dry collected recycled waste which is 1083 thousand tonnes. In the Circular Revolution report, the value of wood waste as a commodity from commercial and industrial, and household sectors was estimated at £2.1 billion. This represented 11 % of the commodity value of waste in these streams, and ranked alongside plastics (13 %) and paper and board (9 %) as significant sectors. Glass—for which there is a well-defined recycling route—ranked as only 1 % of the total commodity value, whereas metals (ferrous, non-ferrous and mixed) accounted for 61 % of the total commodity value (Veolia 2015). In the case of wood, the commodity value may not be easily realised, for reasons which will be discussed later, however, reclaiming the wood for re-use and recycling is clearly worthwhile.

The Wood Recyclers' Association reported the breakout of product types supplied with recycled wood in 2007, based on data from their members and estimated values for non-member organisations (Table 3). This indicated that panel products were 60 % of the market supplied, while the combined total for animal bedding and equine surfaces was 22.4 %, with biomass and energy being only 12.6 %. By 2009, the WRA were reporting that 566 thousand tonnes of the total 2.2 million tonnes of waste wood went into bioenergy (WRA 2010), i.e. the quantity doubled in 2 years. Since then, the total volume of wood recycled has increased, with the WRA reporting that 2.8 million tonnes of material were recycled in 2011, which was 60 % of the estimated total waste wood generated that year (WRA 2015).

2.3.1 Recycling into Wood-Based Panels

It is clear that wood-based panels, such as particleboard present one of the most significant options for recovered wood. Production of wood-based panels in the UK in 2009 was 3.03 million m^3, a decrease from 3.14 million m^3 in 2008, and

3.55 million m^3 in 2007 prior to the recession (FAO 2010). The main products produced in the UK are MDF (660 thousand m^3 in 2009) and particleboard and OSB (2.37 million m^3 in 2009). Of this the input materials were approximately one-third softwood roundwood, one third sawmill residues and one-third recycled wood fibre (from pre- and post-consumer sources). In the years since 2011, this balance has decreased slightly due to the closure of one manufacturer whose primary feedstock was recycled wood, so in 2013 33 % roundwood, 45 % sawmill products and only 22 % recycled wood fibre went into wood-based panel manufacture (Forestry Commission 2014).

UK consumption of panel products is significantly larger than manufacture capacity: 1.10 million m^3 plywood; 1.76 million m^3 particleboard and OSB; 104 thousand m^3 fibreboard and 1.11 million m^3 MDF in 2009 (FAO 2010). It is clear there is scope for recycled wood to occupy a greater proportion of the wood-based panels used in the UK, and also for UK production to increase as the economy recovers from the 2009 recession.

The residence time for panel products in the home is shorter than structural timber. Plywood and OSB may remain for the life of the building, like the softwood joists, as they are structural elements within the fabric of the building. On the other hand, particleboard and MDF are predominantly used in fitted kitchens and other furniture, which have a shorter retention period. It is estimated that the average household replaces their fitted kitchen every 15 years (Couret et al. 2015). For some items, such as flat pack wardrobes, the product life is estimated as being considerably shorter, possibly as low as 5 years (Iritani et al. 2015). Options for recycling of the post-consumer particleboard and MDF products are therefore required to provide an additional life prior to final disposal.

The dominant market for recycling wood waste in the UK is manufacture of panel products—both particleboard and MDF may contain industrial waste wood residues and post-consumer wood collected by local authorities. For the industrial residues, such as planer shavings and sawdust, this material may never be declared as waste, but sold as a co-product direct to the panel mill. As a result, the quantities of recycled material reported by the Forestry Commission in Forest Statistics differ from the quantities recorded as waste by DEFRA. For example, the 2009 production of 3.03 million m^3 panel products, utilised 3.2 green tonnes of wood raw material. 1 million tonnes of this was softwood Roundwood; 1 thousand tonnes was hardwood Roundwood; in addition 1169 thousand green tonnes of sawmill products and 1065 thousand tonnes of recycled wood and fibre were also consumed. The recycled wood and fibre category of panel mill intake is considerably smaller than the sum of the available material reported as waste wood in the UK (WRAP 2009, Fig. 10) where 4.11 million m^3 of timber are included in the municipal, construction, demolition and packaging waste categories. Much timber waste is not recycled into a new generation of products, but proceeds to other applications such as animal bedding or to biomass energy production. A significant quantity is reported to go to landfill, for example if the 40 % of waste wood not recycled in the UK in 2011 was all landfilled, this would be 1.86 thousand tonnes.

2.3.2 Reuse of Wood Waste for Energy or Heat

The other primary option for disposal of waste wood is burning for energy recovery. There is increasing emphasis on renewable energy generation within the UK Renewable Energy Strategy, and wood fuel is expected to contribute significantly in electricity generation but also in heat generation and transport. The capacity for incinerating woody biomass has increased significantly over the past decade, with 134 permitted sites handling 7.5 million tonnes of material (DEFRA 2015). In 2009, it is reported that 560 thousand tonnes of wood residues were recycled into energy generation (WRA 2010). It is known that 346 thousand of this was from sawmills, and 45 thousand from fencing manufacturers (Forestry Commission 2014), leaving the balance of 169 thousand tonnes potentially being wood waste from municipal, or commercial and industrial, or C&D sources. Wood sales to bioenergy from sawmills, and fencing manufacturers, have increased over subsequent years, being 542 thousand and 60 thousand tonnes, respectively in 2013. Over the five years to 2014 it was expected that wood fuel could save up to 7.3 million tonnes of CO_2 emissions by substitution for fossil fuels (Read et al. 2009). Much of this expansion in capacity was in the co-firing market, which increased by 150 % between 2004 and 2009, driven by to the introduction of Renewable Obligation Certificates (ROCs) (see Sect. 3.3).

Many energy generation systems can handle only clean wood, or clean agri-crop residues such as straw, miscanthus and short rotation coppice. For example only 77 of the licenced sites mentioned above can handle waste, the other 57 will be limited to crops and wood co-products rather than any post-consumer or C&D wastes. There is therefore some transferability between wood and agri-crop biomass within the reported capacity, for example utilising miscanthus during the winter and spring but wood waste at times further from the harvest period. Other incinerators operate with greater controls and measures to control emissions from burning timber containing preservative treatments, paints or other potentially problematic components. Compliance with the Waste Incineration Directive (WID, see Sect. 3.4) is required for handling and utilising these materials. The number of available facilities has increased in the UK, for example in 2013, 10.8 million tonnes of waste incineration capacity related to municipal solid waste, commercial and industrial waste, wood waste incineration, gasification or pyrolysis (WRAP 2013).

However, reliance on burning waste wood for energy recovery may be premature, in that it cuts short the options for materials cascading through multiple uses prior to eventual incineration. The waste hierarchy (Fig. 12) places options such as reuse and recycling higher than incineration as strategic goals in waste management.

2.3.3 Landfill Versus Cascading Use

Finally, landfill is the default option in the UK for waste wood which is contaminated, or for waste wood which is too complex a mixture for sorting and

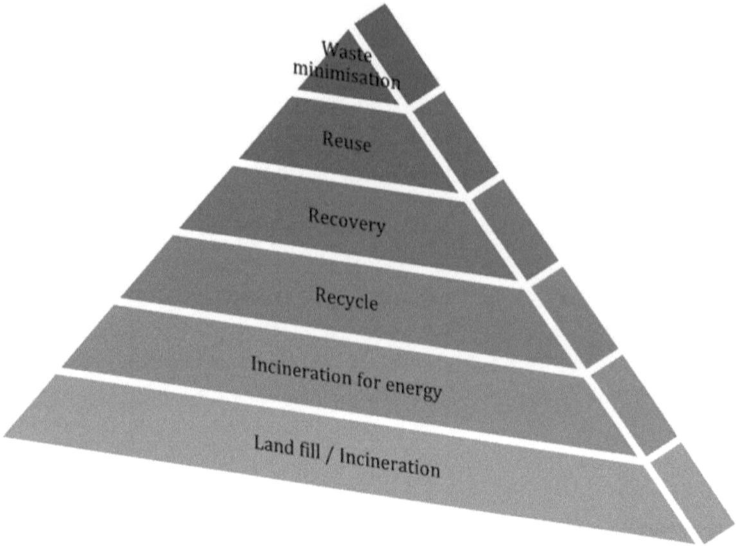

Fig. 12 The waste hierarchy can be seen as a pyramid

segregating. However, it is widely recognised that landfill site capacity is diminishing, and the UK Government has increased landfill tax substantially between 2011 and 2014 to increase efforts to divert materials to other tiers of the waste hierarchy (Fig. 12). Unfortunately landfill may also be the default option for wood where local authority collections are too remote from a suitable processing facility. In the WRAP study, clear regional trends were visible in the levels of recycling which reflected proximity to panel mills (North East, North West and North Wales, South West and Central Scotland). Similar regional trends are likely to apply to incineration as well as to the minority markets such as animal bedding and wood-plastic composite (WPC) manufacture.

The aim of the waste reduction measures, and the philosophy of cascading use of materials, is to reduce the quantity of waste reaching landfill. An ideal materials cascade for waste wood may be to move from primary use in structural timber, pallets, packing, etc. into a second use within particleboard; thereafter the particleboard may be recycled once or more into new generations of particleboard; finally leading to a tertiary product or use for energy recovery. During this process some losses will occur at each step, e.g. non-recoverable off-cuts of particleboard, painted material and material too degraded for recycling, most of these small portions will reach landfill, some may decompose in the environment and a few pieces may also reach new artisan or alternative applications.

Innovative thinking on the recovery and recycling of materials can lead to materials 'cascading' through many life cycles before the material is rendered only fit for incineration and (hopefully) energy recovery (Hill 2011). Figure 13 shows an example materials cascade from the pulp and paper industry. It can be clearly seen

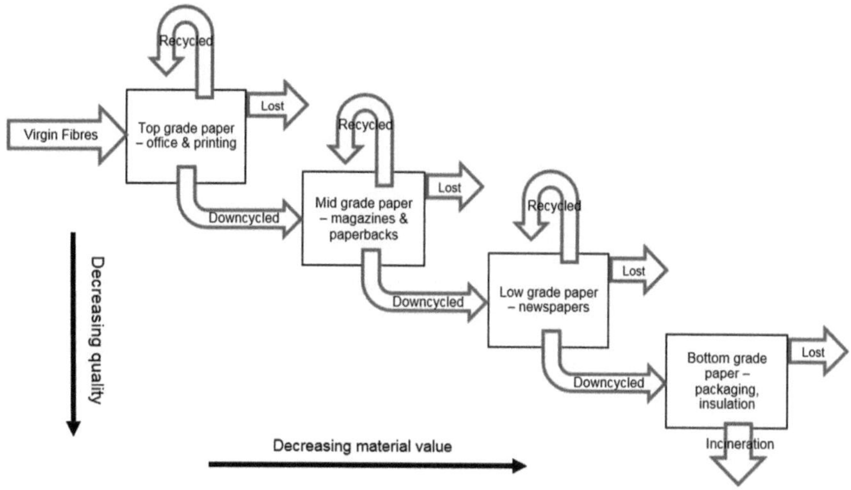

Fig. 13 Cascade of material through multiple products within the pulp and paper industry—an example of the cascading use of materials (adapted from Hill 2011)

that the fibre from high quality paper can be cascaded through several product life cycles before finally being used for energy recovery. In the case of structural timber, it is possible to move wood through different product types down the value chain, prolonging the time period for which the sequestered atmospheric carbon remains in solid form, removed from the atmosphere (Hill et al. 2015).

Within this scheme it is helpful to begin to model the lifespan of a panel product stochastically, attributing a profile for expected lifespan, with for example 25 % to waste by year 1 and 97 % to waste by year 45 as was used for a study of MDF (Medium Density Fibreboard) conducted by Mitchell and Stevens (2009). For a batch of MDF produced in the year 2000, we could for example assume that by 2015 88 % has been disposed of, and only 12 % remains in service. It is then possible to assign probability values for the recycling into new generations of MDF; material directed to energy recovery; and material sent to landfill for each year incrementally. Or, for known annual MDF production levels, the increment from each year's production predicted to reach recovery or recycling in 2015 can be estimated, and suitable recycling, energy recovery and landfill probabilities can be assigned. Such a simulation was used by Couret et al. (2015) to estimate the potential volumes of MDF waste available for nanocellulose production. They hypothesised that in 2012 the volume of MDF waste globally was 58.7 million m^3, while in 2013 this would be 63.8 million m^3. However, stochastic models taking into account a greater range of quality estimations (and thus, suitability for recycling into specific products), in addition to quantity estimations, will be required if the recycling and repeated recycling of particleboard or MDF are to be well understood. For example, on the level that the flows of fibres and quality information within the paper industry are understood.

For waste wood in the UK, two additional potential markets are found in agriculture (bedding for livestock or for equine applications such as arenas and gallops) and in horticulture as mulch. Other minority products such as cat litter and use within WPC products exist but in minor quantities. For most of these minor uses the wood leaves the construction sector, and progresses relatively rapidly to landfill, composting or degradation in service.

2.4 Grading and Use of Graded Demolition Waste Wood

Whilst it is often stated that the separation technology is well established for demolition waste, the separation of waste wood into different categories is much more difficult without provenance and history being known. The Wood Recyclers Association within the UK has categorised wood waste into four main categories (BSI 2012; WRAP 2012).

Category A Category A wood is clean recycled wood. The main sources for this type of wood are the distribution industry (i.e. pallets) retailing, and packaging. The wood can be contaminated with nails, screws and plastics, however, the processors generally screen these out. The wood can also be contaminated with surface paint, however these are usually water based and deemed to be non-toxic.

Category B Category B wood is a mix of Category A and demolition waste. The mix could be up to 60 % Category A wood however, the other material can consist of wood materials from the demolition of buildings and from solid wood furniture. This grade of wood can be contaminated with plastics paints, grit and glues. The major outlet for this type of wood is the particleboard industry and therefore it maintains a value.

Category C Category C wood contains Category B wood, along with fencing products, composite board DIY products and furniture products. As well as the contamination from above, treated wood is also being seen in the mix, this can be in the form of timber treated with newer water-based preservatives and quaternary ammonium compounds. Due to the presence of the preservatives the waste is subject to waste management requirements, and when burnt, needs to be burnt in a WID compliant boiler.

Category D Category D wood includes Category C and track works, fencing and transmission poles. The wood can contain all the contamination found in Category C but also wood contaminated with CCA (Copper Chrome and Arsenic) preservatives and creosote. Due to the presence of CCA and creosote the wood has to be treated as hazardous waste and burned in a hazardous waste incinerator.

Demolition wood features quite high up the waste wood hierarchy, in Category B, however, this relates to sorted clean demolition wood. Strict limits are set, via the WID on the amount of coated and treated timber that is allowable in the timber mix. If the amount of treated timber exceeds *or is likely to exceed* the limits on preservatives and coatings, the timber is down-graded to Category D timber, i.e. hazardous waste which requires specialist disposal facilities.

3 Legislation and End of Life for Timber in the Construction Industry

3.1 Introduction

Government legislation and industry wide adopted standards (whether BS EN standards or industry codes of practice) play a significant role in the determination of the recycling and reuse routes that timber can take. Legislation and policy can be chosen to exert either a pull effect (attracting material into an option by incentives) or a push effect (closing or restricting certain routes).

Legislation has a large influence on the use and repurposing of wood from demolition sites. The process of waste reduction has occurred in several stages, with incremental targets in new generations of legislation. This has occurred in parallel with recognition of the need for greater data to assist decision making. Each step has involved the gathering of additional baseline data necessary for subsequent policy decisions and requirements. For example, whilst the UK Government's review of the Waste Policy (2011) included measures towards creating a zero waste economy, and thus, the reuse and recycling of timber from buildings and structures, the use of demolition wood, because of its potential unknown history, was initially strictly limited. Studies by WRAP and BRE provided greater profile information for waste composition and volumes (CRWP 2009), allowing baselines to be set in 2008 against which future progress in timber and other materials recycling from demolition waste could be measured.

Whilst this chapter does not set out to review all the legislation that has an influence on the use of demolition wood across Europe and beyond, it does illustrate the varied depth of legislation, from Europe-wide directives to Industry standards.

3.2 Waste Framework Directive

C&D waste accounts for almost 30 % of all waste generated within the EU and consist of a mixture of bricks, gypsum, plastics, solvents, asbestos and wood. C&D waste has been noted as a priority waste stream within the EU and high potential for recycling and reuse has been noted with some of the components of the mixed waste stream being of high value. One of the objectives of the Waste Framework Directive (2008/98/EC) is to create a framework that helps the EU become a society with a high level of resource efficiency across all industries. Within the WFD, Article 11.2 states that "Member States shall take the necessary measures designed to achieve that by 2020 a minimum of 70 % (by weight) of non-hazardous C&D waste excluding naturally occurring material defined in category 17 05 04 in the List of Wastes shall be prepared for re-use, recycled or undergo other material recovery". However, it should be noted that this target is for all C&D waste and not specifically demolition timber. Whilst the WFD has set the targets for the reuse of demolition waste the actual reuse across the European Union varies vastly from less than 10 % to over 90 % and it has been noted that if the waste is not separated at source small amounts of hazardous materials (including some preservative treated timber) may make the recycling of demolition waste costly in terms of both time and capital.

3.3 Renewable Energy Directive

The Government of the UK has committed meeting 15 % of the UK's energy demand from renewable resources by 2020, this is a requirement under the European Renewable Energy Directive (European Commision 2009). This was the culmination of several developments in the energy sector, which has undergone a steady transition of energy balance during the past 20 years. The target of 15 % was achieved in 2013, and renewable energy output has continued to increase.

ROCs were introduced in 2002 in England, Wales and Scotland. This created a market incentive to achieve the policy target, and allowed suppliers generating energy from renewables, including biomass and co-firing of biomass, to trade the certificates with suppliers whose energy generation was as a result of non-renewable means. This lead to increased investment and a trend to increase total energy generation from renewable sources. The Renewables Obligation was reviewed, and additional measures were introduced in 2011. The number of energy from waste plants in the UK has increased, and additional capacity is planned. These factors have changed the value of certain grades of waste wood, where sufficient information is known for it to be sold to the energy industry as biomass rather than waste. The new incineration capacity has made biomass energy an economically viable option in new regions, although recycled wood uptake in the panel products sector remains high where this has been a traditional recycling option.

The ROCs had a significant impact on the energy industry in the UK. Energy from plant biomass reported by DEFRA (2015) was 8933 GWh in 2013 (Fig. 14), this showed exponential growth of the biomass energy sector over a 5-year period. The incineration of waste for electricity showed more modest growth over the same period, reaching 1987 GWh. Waste incineration remains relevant to the fraction of wood waste which is suspected to contain paints and preservatives.

The UK Renewable Energy Strategy committed to the 15 % renewable energy target by 2020, it also indicated that wood fuel would make a significant contribution to electricity, transport and heat generation sectors (HM Government 2009). The use of biomass for energy in the UK is therefore increasing, and the use of waste wood arisings is becoming a more commonplace substitute for virgin biomass. This is a significant market for waste wood, however, the industry is

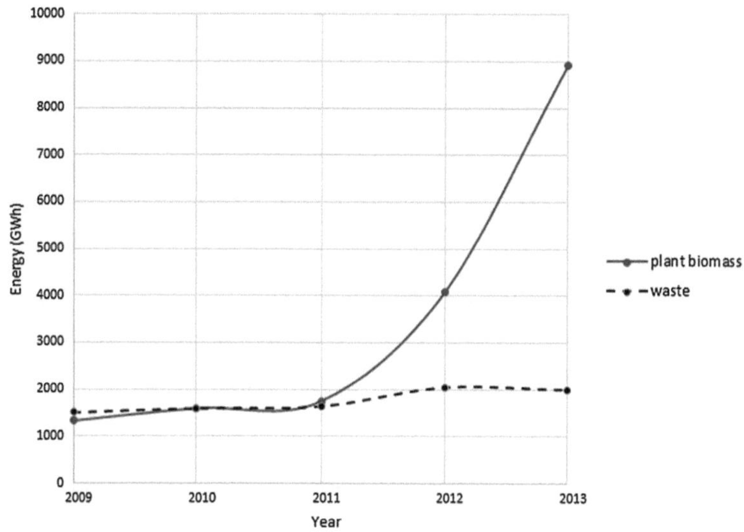

Fig. 14 Electricity generated in the UK from plant biomass and from waste 2009–2013 (DEFRA 2015)

regulated both in terms of targets set by National Renewable Energy Delivery Plans, and associated delivery incentives, and in practical terms by the WID, governing which categories of timber can be converted to energy under what conditions.

3.4 Waste Incineration Directive (WID)

The aim of the WID (EC Directive 2000/76/EC) is to prevent or limit the negative effects on the environment of incineration or co-incineration of waste (DEFRA 2010). It includes measures to prevent or limit as far as practicable, pollution of air, soil, surface water and groundwater. It includes measures such as flue gas cleaning plants and minimum requirement conditions for combustion within the chamber. If wood waste is clean (i.e. without the presence of paints, halogenated treatment agents or heavy metal-based treatments) then it is excluded from the WID requirements, however, wood waste from mixed sources, including any of these treatments, must be handled in a WID compliant facility. WID was recast into the Industrial Emissions Directive (2010/75/EU) in December 2010, and became law in the UK in 2013. The IED made very little change to the WID requirements, however, small alterations of control over pyrolysis plants and gasification plants allowed process-relevant conditions to be used as long as the gases produced were purified to such an extent that they were no longer waste prior to incineration, and caused emissions which were no higher than the burning of natural gas.

Whilst incineration plants utilising only wood are generally exempt from needing a WID permit, plants that utilise treated timber that may contain halogenated compounds or heavy metals, will need to be regulated and permitted. The utilisation of wood products in incineration plants is also commented upon in the WID. Whilst the Directive acknowledges that halogenated compounds and heavy metals are not utilised in the manufacture of wood-based panel products, it suggests that the panels could be manufactured from recycled wood that does contain these chemicals. Therefore, the onus is on the incinerator operator to prove that the panel products contain no treated wood in the making of the boards. It is interesting to note that whilst the Wood Recyclers Association separates demolition wood and contaminated wood, the WID envisages that wood arisings from C&D activities are likely to have been treated in some way. Unfortunately, this implies that all C&D waste wood is covered by the Directive so a permit will be needed to burn all C&D wood. It may be possible for the producer of C&D waste to sort and segregate the wood waste if technology permits, and declare end of waste for the clean fraction. It is also the case that if wood enters a waste transfer station it is again the responsibility of the incinerator operator to prove that the wood in the transfer centre has not been contaminated with treated timber.

If an operator wants to use treated wood in their incinerator, a permit will have to be applied for before operation. The permits will only be granted under strict requirements, which include the types of waste being used, the delivery areas of the incinerators, the type of combustion furnaces, the abatement plants, residue handling and monitoring equipment employed at the plant. The permit will also set out the emission limits that have to be adhered to by the operator.

3.5 Landfill Directive

The Landfill directive (1999/31/EC) is a European directive aimed at the prevention or reduction of the negative effects on the environment, in particular the pollution of surface water, ground water, soil and air. It also covers the global environment, including greenhouse effect, as well as any risk to human health from landfilling waste, during the whole life cycle of the landfill. The directive is applicable to all types of landfill, from hazardous waste to inert sites. It is also noted in the directive that EU member states should reduce the amount of biodegradable municipal waste from 1995 levels by 50 % in 2009 and to 35 % of the 1995 values by 2016. It is interesting to note that the disposal of timber is effected by the two main guiding principles of the directive. First, the prevention of hazardous waste being disposed of in non-appropriate sites—the classification of preservative treated timber as hazardous waste means that treated timber will have to be disposed of in specially designated hazardous waste sites and at a high disposal cost. Second, the fact that timber is a biological material means that the landfilling of timber is seen as the landfilling of biological waste, which also has reduction targets.

Generally, countries within the EU use landfill taxes to achieve their targets as laid down by the EU. Current landfill tax in the UK (April 2015) is set at £82.60 per tonne (HM Revenue and Customs 2015). This value shows an increase of one order of magnitude since the tax was first introduced in 1996.

Whilst the amount of timber entering landfill from the C&D sector has fallen in real terms (i.e. in the tonnage entering the landfill) since the introduction of the landfill directive to present day, the composition of the total waste from a demolition site has changed dramatically. A steady reduction of the weight of inert C&D waste has occurred, due to the increase in reuse and recycling of aggregates on site, which are the result of the first wave of waste reduction targets for the industry. In 2008, it was estimated that the inert materials comprised 76.5 % of the total landfilled material from C&D, while timber was only 0.3 % of the total. However, this represented 13,828 tonnes of landfilled wood from C&D waste, with the destination of a further 61,420 tonnes unknown (WRAP 2008). It is now important to address this more visible timber waste stream, and the challenges associated with classification of the waste as hazardous relating to the unknown levels of treatment products which it may contain.

Timber will continue to be landfilled until a viable option is found for the handling of preservative treated waste timber, whether this is a technology for the remediation of the preservatives from the timber before reuse or a viable incineration plant with energy recovery for the conversion of contaminated timber to energy.

3.6 *Industry Adopted Regulations*

Many industries (and even individual companies) will specify the quality of the waste wood being utilised in their products. The specification may range from the size and form of the timber for a specific use, to the permissible levels of contamination within the timber. One industry that has come together to develop an industry standard is the wood panels industry.

The panels industry (and particularly the particleboard industry) utilises a significant amount of waste wood in their manufacturing process, however, the wood has to be clean and free from chemical contamination before it enters the factory.

The Wood Panel Industries Federation (WPIF) recognised the use of waste wood in the manufacture of panel products in 2000. The WPIF developed an industry standard (WPIF/UKFPA/1-2000) which was based around the strictest European Standard of the time for allowable metals content, EN 71 'The Toy Standard' (BSI 1993). This standard prescribed the heavy metal content limits for materials used in the manufacture of toys. This was followed in 2004 by a Publicly Available Specification (PAS 104, BSI 2004) to govern contaminant levels, with an updated WPIF guidance document following in 2005. WPIF/IG/12.2005 'Guidance for the specification and control of post-consumer reclaimed wood raw materials used in the manufacture in the UK of wood particleboard, MDF and OSB' was the resulting

Table 4 The maximum allowable quantities of contaminants permitted in particleboard manufactured to industry standard WPIF/UKFPA/1-2000

Contaminant	Limit (mg/kg)
Arsenic (As)	25
Cadmium (Cd)	50
Chromium (Cr)	25
Copper (Cu)	40
Lead (Pb)	90
Mercury (Hg)	25
Fluorine (F)	100
Chlorine (Cl)	1000
PCP	5
Creosote	0.5

standard for wood feedstocks to the panel products sector. The WPIF standard was also adopted by the European Panels Federation (EPF 2002).

Table 4 details the limits prescribed in the industry standards. These limits have been incorporated into the more recent PAS 111 standard for processing waste wood, with an additional restriction on the total heavy metal compound content of 4000 mg/kg dry matter (BSI 2012). Whilst the standard was written to protect the panel board industry's interests, and ensure that panels were able to supply to all industries, it prevented the use of most demolition waste wood streams within the sector and instigated the sector's dependence on clean waste wood. Much work has been undertaken to develop methods for the analysis of timber waste streams, in the first instance as quality control of waste wood into the industry, however with the ultimate aim of segregation and decontamination of timber prior to it entering the panel manufacturers.

3.7 *Infrastructure Effects*

The introduction of site waste management plans (SWMPs) for construction projects of £300,000 or more from 2008 onwards has boosted activity to segregate wood waste at source. It has also altered the quantity of data available for this sector, as the company has a responsibility to estimate the types of waste which will be produced prior to any activity; then to record waste generated during the process by type and quantity. This has led to simple steps to increase segregation of materials on site (Fig. 15). One of the aims of the introduction of SWMPs was to increase environmental awareness of both the workforce and the management. It has also increased profitability and efficiency due to greater knowledge of the resources present and wasted on site.

Total waste in the UK is reported by DEFRA (2015). This is grouped by sector, with C&D generating 100 million tonnes in 2012; commercial and industrial generating near 50 million tonnes, and household waste generating 26.5 million tonnes in 2012. The C&D data has been stable since the start of the recession,

Fig. 15 Timber removed during construction work may be a mixture of structural material from the building, and site-generated waste such as hoardings, covers and formwork timbers

whereas the commercial and industrial, and the household sectors have decreased. Timber is not segregated in the overview data, but in construction is likely to have remained at a consistent percentage of the total. In WRAP (2009) the construction sector wood waste was 0.89 million tonnes, formwork waste was 0.29 million tonnes, and demolition timber waste 0.81 million tonnes.

The commercial and industrial sector included waste from the manufacture of wood and wood products, which was quantified as 2.047 million tonnes in 2009, 2.193, 2.264 and 2.160 million tonnes in 2010, 2011 and 2012 (DEFRA 2015). This figure is four times larger than the segment calculated by WRAP in their 2009 report for industrial wood waste (462.5 thousand tonnes), however, it may reflect differences in reporting methodologies. DEFRA reported that 'Furniture, other manufacturing and repair' generated a further quarter of a million tonnes, however, this may include upholstery, foams, plastics and other materials in addition to some wood offcuts.

Downstream, the wood-based panel manufacturers reported an intake of 1.169 million green tonnes of sawmill co-products and 1.065 million green tonnes of recycled wood in 2009, in addition to 1.009 million green tonnes of softwood roundwood (Forestry Commission 2010). The sawmill co-products: waney edge boards, small dimension logs and chips, plannings or sawdust, are likely to contribute to the gross figure reported by DEFRA (2.160 million tonnes), although these co-products are readily utilised. The recycled wood intake to panel manufacture is likely to come from municipal waste (depending on location), pallet and packaging waste, and post industrial waste (joinery and furniture manufacture). The pallet and packaging waste provides a ready supply of preservative free feedstock suited to panel production.

3.8 Summary

Whilst this section has not shown an exhaustive list of legislation and standards that govern the use of demolition wood it has shown the breadth of legislation, and some of the context for use of waste wood. This includes directives that govern what happens to waste timber on site; incentives to increase incineration to meet renewable energy targets, but restrictions on whether waste wood can be burnt; and targets to reduce volume to landfill. In addition, the potential to re-use waste wood to make new products will be considered later in this chapter.

Whilst it appears that clean wood has a route to market supported by legislation, contaminated or potentially contaminated timber has a much more limited market. This market will become more restricted as legislation becomes tighter, yet a large volume of C&D wood will be diverted to recycling facilities and enter the system with insufficient documentation to allow it to be graded as Category A material. There is considerable scope for improved sorting technologies, for example the use of spectroscopy (Morak et al. 1999; Moskal and Hahn 2002) or machine sorting systems (Moskal and Hahn 2002; Hasan et al. 2011) to identify preservative treated wood. However, such systems may be suited for only certain contaminants (e.g. copper or arsenic). On the other hand, new products and processes to deal with treated timber are being currently researched, as will be discussed in Sect. 5.

4 LCA in the Construction Sector

4.1 LCA for Timber in Construction

Central to understanding the environmental performance of wood within the construction industry is a robust and transparent environmental assessment method. While a range of such methods exist, Life Cycle Assessment (LCA) has become a de facto approach, due to the relative comprehensiveness of the environmental impacts it assesses. Unlike single-issue methods such as carbon, water or ecological footprinting, LCA quantifies a broader range of environmental indicators spanning ecological degradation, resource depletion and human health. In doing so, it avoids the danger of ‘burden shifting’, whereby a product is selected based on an advantageous profile in one area (e.g. a lower carbon footprint) at the expense of an unforeseen effect in another. LCA is therefore becoming more widely used within the sector, in particular as schemes such as BREEAM and LEED promote sustainable practice in construction. While its use at this stage remains primarily limited to the more advanced economies (as it is with LCA more widely), this is likely to change as it becomes more embedded within the sector and more geographically explicit life cycle inventory (LCI) data becomes available for use.

At the primary usage stage (i.e. the first service life) most of the commonly used wood-based construction materials have been well studied by the research community. Peer-reviewed data exists for wooden-framed building structures of various configurations (e.g. Gustavsson et al. 2010; Ximenes and Grant 2013), wooden floor coverings (Nebel et al. 2006), window frames (Asif et al. 2002) and hard-boards (González-García et al. 2009). Studies of panel products exist for orientated strand board (e.g. Kline 2004), LVL (Wilson and Dancer 2004), particleboard (Rivela et al. 2006) and MDF (Rivela et al. 2007). There is also data for a number of more complex wood-based products, such as composite I-joists (Puettmann and Wilson 2013) and a pre-fabricated ventilated wall unit (González-García et al. 2012). Data for glulam production is available through the Ecoinvent database (Athena Institute 2012; Weidema et al. 2013). Where end-of-life is modelled, these studies tend to assume incineration with energy recovery as the most likely fate, with a number acknowledging the potential for further reuse and the environmental benefit that may be achievable through doing so (though without explicitly developing this).

In its more natural state (i.e. as relatively unprocessed timber for frames, beams, joists etc.) wood has a demonstrably less harmful environmental profile than a range of other common building materials. A number of studies have shown the preferential performance of wooden-framed structures over concrete (e.g. Börjesson and Gustavsson 2000; Lenzen and Treloar 2002) and steel (John et al. 2009) in terms of their embedded energy and greenhouse gas (GHG) emissions. Ximenes and Grant (2013) demonstrated that their ‘wood maximised’ house—a structure comprising timber elements in all major structural aspects of the building—produced 50 % less GHG emissions over its lifetime than a market-typical control. Wood in this state requires comparatively little energy to process and the energy that is used is typically generated from incineration of by-products. In the EU, this typically constitutes 69–83 % of the total primary energy demand during manufacture (Bribián et al. 2011) and this replaces fossil-based energy.

As wood becomes more refined or processed, the embodied energy demand increases, as does the impact of additives on the LCA results. Synthetic resins, adhesives, preservatives and finishes all increase the environmental burden of wood-based products, in particular in relation to toxicity-based indicators. While some additives are integral to the formation of the product (e.g. adhesive resins in

MDF; preservatives in treated timber) others are secondary products interlinked with the wood in its use-phase (e.g. paint, varnish or oil on a door frame). Results for timber floor coverings, for example, are highly sensitive to assumptions made about the glues used to fix them and any finishes used once they're in place (Nebel et al. 2006). The development of more environmentally benign additives is a focus of research and has the potential to reduce the footprints of processed wood-based products further. For example, it is estimated that replacing urea-formaldehyde and melamine-formaldehyde with bio-derived adhesives could reduce CO_2 emissions associated with laminated woods by up to 16 %, and with fibreboards by up to 46 % (Bribián et al. 2011).

In spite of the observations above, there is, perhaps inevitably, considerable variation among the output of LCA studies within this sector. This is due only in part to variation in the scope of individual studies and the differences in functional unit, system boundaries, modelling approaches, etc. that this entails. Key assumptions made about the use and performance of the materials themselves also play an important role in this variation.

One criterion about which there is considerable uncertainty is that of the assumed service life of buildings. In the context of LCA, this assumption plays a crucial role in the final output of the analysis. A building assumed to have a lifetime of 50 years has a *functional equivalence* of half a similar building assumed to last 100 years. In other words, it will need to perform twice as well (in all non-use phases of the lifecycle) in order to have an equivalent environmental performance over time. However, given the long timescales involved in the use-phase of most buildings, service lives in relation to construction are often chosen based on typical values rather than by direct assessment (Aktas and Bilec 2012; Rincón et al. 2013). These values can show considerable variation (see Table 5) and may not adequately

Table 5 Variation among assumed and calculated service lives of buildings

Source	Analysis	Service life
Bribián et al. (2011)	Review of 60 studies across 9 countries	Typical service life = 50 years, range = 30–100 years Large variation across countries: Netherlands: 75 years, residential; 30 years, offices UK: 60 years, both residential and commercial Finland: 100 years Switzerland: 80 years
Aktas and Bilec (2012)	Review of 17 papers covering residential buildings	50–100 years
	Quantitative assessment of US residential buildings. Based on data from US Census Bureau	61 years, within range of 21–105 years with 90 % confidence
Rincón et al. (2013)	Quantitative assessment of Spanish residential buildings, based on INE (Nacional Institute of Statistics) data	National average = 80 years Significant variation across regions: Low = Cueta, (54 years) High = La Rioja (95 years)

reflect reality. This is especially so when in-life factors such as maintenance regimes (whether adequate or sub-standard), and changing consumer tastes and demands over time, are taken into account. Models that can better approximate building service lives, especially with respect to consumer behaviour, would be helpful here, especially since better reuse and recycling offers the potential for multiple service lives to be realised from each piece of wood.

The choice of functional unit is an essential factor when modelling construction scenarios, and can have significant influence. Wood-based materials may have subtle secondary roles within the building, whether these are structural, protective, insulative or ventilatory. Thermal energy for in-life heating, in particular, can play an important role in LCA results of whole buildings so materials that play a secondary role in thermal insulation will need to be considered with this in mind. This is especially relevant in colder climates, where lifetime embodied energy is heavily dominated by heating during the use phase (e.g. Adalberth 2000).

Another area of particular relevance to wood-based products, especially within the context of long-lived building structures, is that of stored carbon. By removing timber from sustainably managed forests (i.e. forests where the timber extracted each year does not exceed the quantity regrown in a year) and using it in long-lasting building components, wood has the potential to increase the store of carbon within the building stock without reducing the carbon sequestered in forests (Hill and Norton 2013). The total amount of stored carbon in the UK housing sector is estimated to be 19 million tonnes, which is equivalent to 70 million tonnes CO_2 (Read et al. 2009). The carbon sequestration and storage benefit is further extended whenever subsequent service lives are established to prolong the useable lifetime of the biomass. However, in terms of carbon accounting protocols (at the product level), the most widely used standards, PAS 2050 (BSI 2011) and GHG-Protocol (WBCSD/WRI 2011), along with the ILCD (European Commission 2010a) stipulate that carbon must be stored for a minimum 100-year period before it can be factored directly into the calculation. While this allows for the inclusion of carbon stored in structural timbers in the existing Victorian housing stock (now well over the 100 year threshold), many modern building materials, with their assumed shorter service lives, miss out. This remains an area of debate for wood-based products, though the relatively new standard for construction sector Environmental Product Declarations (EPDs), EN15804 (BSI 2013) takes the same stance as the approaches outlined above.

This scope for variation in LCA can frustrate commercial stakeholders who seek clear, readily accessible answers when considering which materials to use for a particular project. While construction professionals typically recognise the advantages of LCA, they cite its complexity, along with time and data demands as barriers to its further uptake (e.g. Olinzock et al. 2015). In view of this, there is recognition of the need for a more harmonised approach to product environmental assessments, in particular with a view towards improving the inter-comparability of results. To this end, initiatives such as the EU's Product Environmental Footprint

(PEF) project and a range of EPD schemes (e.g. see Passer et al. 2015) have sought to develop a standardised approach to assessing environmental profiles. More instructive than the ISO14040/44 standards, these schemes use sector-specific Product Category Rules (PCRs) to outline how products should be assessed for a given market sector.

In the case of construction, EN 15804 provides rules specific to the preparation of EPDs for all construction products and services. The standard cites seven impact factors for inclusion (global warming, ozone depletion, acidification of land and water, eutrophication, photochemical ozone creation, abiotic depletion of elements, abiotic depletion of fossil resources) and splits the lifecycle into four modules (see Table 6). Waste wood can occur in any of these modules, for example through manufacturing and construction waste (module A), in-life repair and refurbishment (module B), as well as final demolition and disposal (module C). Instructions are provided for the standard calculation of LCAs across the sector, with a view to ensuring the inter-comparability of results. However, some have already highlighted differences between EN 15804 and the approach taken in PEF, in particular in terms of the impact factors included (PEF outlines 14 impact factors, for example) and the approach to modelling recycling (Passer et al. 2015). It remains to be seen how this will play out (at the time of writing, the PEF programme is still in development) but greater harmonisation of approach remains both a goal and a challenge in terms of further establishing environmental declarations within business and policy-making sectors.

Table 6 Lifecycle modules outlined in EN15804 from BSI (2013)

Module		Sub-stage
A	Product stage	A1—Raw material supply A2—Transport A3—Manufacturing
	Construction process	A4—Transport A5—Construction/installation
B	Use	B1—Use B2—Maintenance B3—Repair B4—Replacement B5—Refurbishment B6—Operational energy use B7—Operational water use
C	End-of-life	C1—Deconstruction/demolition C2—Transport C3—Waste processing C4—Disposal
D	Benefits and burdens beyond the system boundary	Reuse, recovery and recycling potential

4.2 LCA for Current Forward Supply Chains

Construction and wood-based products pose particular challenges in terms of modelling end-of-life options in LCA. In part this is due to the long service lives of buildings and the uncertainty that this brings to end-of-life scenarios that may be 50-100 years from realisation. How will disposal and reclamation technologies have changed when the material reaches its end-of-life, how much more efficient will our existing options have become, and what new technologies may have become available by the time end-of-life is reached?

Closer to home, there is significant uncertainty (or at least, scope for variation) about the end-of-life options currently available, in particular in terms of performance of wood in landfills and the thermal efficiency of indirectly modelled energy recovery scenarios. Furthermore, these uncertainties are significant because the results of LCA in relation to wood-based products are very sensitive to assumptions made about end-of-life. Werner and Richter (2007), for example, highlighted the sensitivity of LCA output to both assumptions made about emissions of methane from wood in landfills and the thermal energy recovered at incineration. Sandin et al. (2014) went further in exploring the influence of end-of-life assumptions. Their study tested a range of EOL scenarios including sensitivity to the energy source used during demolition (renewable v diesel); fuel used for transportation (biodiesel v diesel); disposal route (incineration v recycling); and approach to modelling allocation at recycling (i.e. a cut-off approach or consequential substitution—see below). The analysis was conducted in the context of a comparison between glulam and steel roofing beams using average European data for the end-of-life options. Their analysis demonstrated that while the *relative* results between to two products were 'remarkably robust' (glulam outperformed steel in all cases, except where substitution was used to model the recycling scenario, which gave comparable results)—the results varied substantially in absolute terms. The most important assumptions each had a critical bearing on the outcome and were seen as: choice between recycling or incineration; choice of modelling approach for recycling; and the choice of whether to include the end-of-life phase in the study at all (i.e. between a cradle-to-gate or cradle-to-grave system boundary).

Sandin et al. (2014) also tested scenarios relating to future technological development: one in which today's low impact technologies (i.e. renewable energy, bio-based transport fuels) were assumed to have become the norm; and another in which developments in efficiency were such that EOL impacts became negligible when compared against the other lifecycle stages. Again, the results were highly sensitive to these future technologies assumptions and they included this among those factors they considered most critical to successfully modelling EOL in LCA. In conclusion, they recommended making a temporal distinction when conducting LCAs of construction products whereby uncertain future EOL impacts are presented separately to those occurring in the shorter term, and are accompanied by sensitivity analyses. This is especially relevant when assessing new or recent builds, where the end-of-life scenario may not be realised for many years.

The following sections will address the three primary EOL options currently available and with wide uptake across the UK: incineration, landfill and recycling into panel products. Novel approaches within the construction sector and related products will be addressed in Sect. 3, and emerging technologies in which the material is transferred outside the construction sector are addressed in Sect. 4.

4.3 Incineration with Energy Recovery

Incineration with energy recovery is a widely favoured route for disposal of spent wood-based materials, at least within the European context. Arguably an additional service life rather than simply a disposal option (the product in question being energy), this option can render lifecycles of wood product carbon-negative overall. Gustavsson and Sathre (2006), for example, explored a number of scenarios relating to the lifecycle of a wooden building frame (including lumber drying efficiency, transportation distance, carbon intensity of fossil fuel, and use of forestry and processing by-products for biofuel) and concluded that the energy balance was negative in all cases where demolition wood was recovered for energy. Furthermore, they found that energy recovery was the most significant factor in the wooden frames' lower energy and CO_2 balance when compared against concrete.

In spite of this, the dynamics of incineration vary widely depending on a number of factors, including the moisture content of the wood going in, the type and assumed (or measured) efficiency of the energy recovery system being modelled, and the carbon intensity of the fossil energy being replaced. EN 15804 stipulates that energy recovery efficiency must be at least 60 % for any given material for a process to be considered as such, however, there is wide scope for variation above this figure. In terms of modelling, a key factor in the calculation of benefits attributable to energy recovery is the choice of 'avoided product' (i.e. the product assumed to be replaced by the generation of energy from waste). This may be avoided natural gas consumption in the case of heat energy, or avoided consumption from the grid for electricity. However, GHG emissions (to pick one impact factor) associated with the consumption of grid electricity vary widely from country to country. Avoided emissions associated with replacement of grid electricity in India, for example, will be approximately 2.7× greater than the EU average, and in Australia they will be 2.3× higher (DEFRA 2015). Even within Europe there is significant variation, with GHG emissions associated with UK grid electricity consumption approximately 6× higher than those of France, and 29× higher than those of Norway (where the grid mix is dominated by renewables) (Ecoinvent data, Weidema et al. 2013). This means that great care needs to be taken when comparing results of energy recovery studies from different countries, and awareness is needed of the role played by avoided emissions assumptions in LCA results of this form of disposal.

Incineration of Category D waste may present a well regulated option for handling wood waste which is known to contain CCA-treated wood, or other preservative treatment systems. Where suitable boilers and emission control systems exists the incineration for energy route offers energy recovery for the biomass, leading to a favourable global warming potential compared to other incineration feedstocks. The methods for controlling air emissions include the use of prescribed temperatures, scrubbers and sorbent systems (Iida et al. 2004). The use of calcium-based ash in the combustion of waste wood can increase retention of arsenic and chromium, and reduce the leaching of the arsenic and copper. Sodium systems assisted with the retention, however, the metals remained leachable.

4.4 Landfill

While landfill remains the default option for disposal of Category D wood waste in the UK, due to insufficient capacity in WID compliant incinerators, this may not be an optimal solution. Problems are known to occur in unlined landfill sites, where CCA-treated wood within demolition waste can lead to arsenic content in the landfill leachate composition (Weber et al. 2002; Khan et al. 2006). The species of arsenic detected in the leachate varied between the types of landfill studied by Khan et al., with dedicated C&D landfills showing the lowest quantity per metre of water added (0.36 % per m compared to 0.69 % per m for wood mono-fill and 0.84 % per m for municipal solid waste). However, the projected cumulative values for arsenic leached from the 21 Florida landfill sites was estimated at between 20 and 50 tonnes up to the year 2000. A significant increase was expected to 2040, with peak levels at 2100. It is reasonable to question the viability of landfill as a long-term solution for CCA-treated wood.

In a LCA comparing landfill and waste to energy incineration (in Florida) it was found that waste to energy presented a viable alternative (Jambeck et al. 2007). The metal ions retained in the ash or emission control system present a much more concentrated and controllable form than the dilute metal ion levels emitted in a large volume of water from unlined C&D landfill sites. Where possible landfill of waste wood in municipal solid waste sites was preferred, as these are lined, unlike the C&D landfill sites in Florida.

Although landfilling waste is widely considered to be a poor approach to waste disposal, the carbon content in wood, alongside the anaerobic conditions in landfills, means that long-term carbon storage becomes a consideration when disposing of spent wood in this way. LCA results are very sensitive to assumptions about the degree of decomposition of the wood over time (Ximenes and Grant 2013), the methane generated (Werner and Richter 2007), and how that methane is handled (i.e. whether it is captured for energy recovery or not). Ximenes and Grant (2013), for example analysed a number of landfill scenarios for construction wood in Australia and showed that landfills could act either as a store of carbon in wood or as net emitters of carbon, depending on the decomposition factor used. The most beneficial scenario they modelled (in terms of carbon balance) outperformed their incineration with energy recovery scenario and was based on wood in a C&D

landfill. This used a USEPA figure of 24 % decomposition of degradable organic carbon (US EPA 2006). In this scenario, they found landfilling to be approximately 65 % more carbon-efficient than incineration with energy recovery, storing 1.78 tonnes CO_2 per tonne of landfilled timber. They highlight the predominantly inert content entering Australian C&D landfills and the near absence of moisture within them as being factors significantly limiting the onset of anaerobic decay.

Aside from considerations of GHG balances, potential environmental impacts associated with landfilling include contamination of groundwater and soils, through leachate release, and other toxicity-based impact factors associated with the decay of chemically treated woods.

4.5 Recycling

Studies as summarised in the review by Iritani et al. (2015) have demonstrated that recycling wood waste into particleboard has a clear LCA benefit, when compared to other scenarios. The papers cited frequently favoured medium density particleboard over MDF for furniture manufacture, due to the higher recycled content of PB over MDF (e.g. Bovea and Vidal 2004). The ecological footprint of the wood-based panel may also be reduced for products containing recycled material, as energy inputs are frequently lower for this drier feedstock, and less land is required (attributed) for production (Saravia Cortez et al. 2013).

Recycling of demolition wood in the UK is only partially taken up, due to the categorisation of unknown timber which may be treated with preservative chemicals as hazardous, and Category D. In the USA a demonstration study formed flakeboards from recycled CCA treated timber (Vick et al. 1996), with the intention that such products may be suitable for exterior products such as siding, where the preservative treatment may be re-used appropriately. Adhesive performance with CCA treated timber is known to be poorer in some cases than with the untreated materials in common use, however the use of a hydroxymethylated resorcinol primer was shown to overcome this difficulty (Vick et al. 1996). An industry survey conducted in the USA showed that few wood composite manufacturers were in favour of reuse of CCA-treated wood in wood composites, preferring that this material be handled by specialist facilities, used for fuel or used in wood-non-wood composites (Smith and Shiau 1998).

It's important to note that while recycling represents the end-of-life for a product, it does not represent the EOL of the material. In material terms, recycling is not a direct comparison to incineration or landfilling since it generates a further service life, which in turn may end in one of these final disposal routes. When considering LCA at the product level, this point can be lost since the only consideration is how that particular product is disposed of. Because of this material throughput, modelling recycling in LCA can be complex and has resulted in differing approaches depending on where one considers the benefits of recycling should lie. These approaches have various names but are perhaps most succinctly summarised as the 100:0, 0:100 and 50:50 approaches. Since it is known that the LCA results of waste wood utilisation are highly sensitive to allocation procedures (Werner et al. 2007a; Sandin et al. 2014) they are worth further consideration here.

The 100:0 approach, also known as the recycled content method or the cut-off method, allocates the benefits of producing recycled content with the *user* of that recycled material. The original system boundary is cut-off at the point of disposal and the recycled material enters its new lifecycle burden-free, barring those impacts associated with the recycling process itself (i.e. collection, sorting, reprocessing, etc.). The carbon accounting protocol PAS 2050 cites this approach for use in open loop recycling scenarios, i.e. those where the recycled material 'does not maintain the same inherent properties as the virgin material input' (BSI 2011). This would be the case in most, if not all, recycling scenarios involving construction timber and other wood-based products, according to this classification.

The 0:100 approach, also known as closed-loop approximation or recyclability substitution approach, allocates the benefits of recycling to the *producer* of the recycled content (i.e. to the original product lifecycle). This approach requires a system expansion to account for the 'avoided use', at a later date, of virgin material. In PAS 2050, this is typically applied to endlessly recyclable materials such as metals, or where a consequential approach is useful to broader policy-based studies involving wider system boundaries. EN 15804 cites this as the method to be used when assessing construction sector products for EPD purposes. Under this protocol, burdens associated with demolition, transport of waste to a processing site and the waste processing itself are assigned to modules C1–C4 of the lifecycle, while the benefits associated with recycling are accounted for in Module D. EN 15804 highlights the need to ensure functional equivalence when calculating avoided burdens for module D, so where additional processing of recycled material is needed in order to achieve this, it should also to be included this module.

Seen by some as a compromise approach, the 50:50 method splits the benefits of recycling between the producer and user of the recycled content. Although there are a range of ways in which this can be achieved, it has perhaps gained greater prominence since its inclusion in the EU's PEF project, which aims to provide a harmonised approach to product and organisational footprinting. A formula for the calculation is provided in their documentation (EU JRC 2012).

5 Novel End-of-Life Approaches to the Recycling of Timber Within the Construction Sector

It has already been shown that the markets for waste wood from construction are limited due to contamination of the timber through its first tier usage and through the lack of known history of the timber. However, new research is leading to possibilities for the use of waste demolition wood for new applications. The research is leading to novel processes for detection of and cleaning of pollutants from timber that then can become a viable feedstock for new products, or to new products and processes that can utilise polluted timbers with quantified contaminants.

5.1 Reclamation and Reuse of Solid Wood

Whilst the reclamation and reuse of timber is nothing new (timber has been reused throughout history) this practice has become more difficult in many large demolition projects, due to the scale of demolition works, safety considerations and efficiency in clearing a site becoming separate from subsequent construction by another team on a different contract. However, small building projects and conservation projects will continue to generate and use reclaimed timbers, ranging from parquet flooring to windows, stair spindles and other mouldings. Much of this small scale re-use by opportunists can never be fully quantified as it generally occurs outside the standardised reporting mechanisms, however, reclaiming structural timbers, or decorative products such as parquet flooring blocks, presents a significant opportunity. Reclaimed wood may be stored and marketed by salvage merchants, and specialist architectural dealers. Increasingly, the small contractor may also use online catalogues of available or desired materials, and internet sales (Fig. 16). On the other hand challenges exist if this timber is to be used in mainstream construction activities, due to grading requirements or balancing availability against demand at a given location and time. Nonetheless, it is worth significant thought as initiatives to re-use timber rather than recycle it offer the lowest embodied energy, and greatest scope for extending service life, where removal of intact timbers is possible.

The use of reclaimed wood is becoming more prevalent for several reasons. Primarily the use of high quality aged timber is at the moment 'in vogue' and the distressed old look is one that is desired within buildings. The reclaimed wood can be repurposed to many applications, including flooring, furniture, garden and outdoor applications. However, there is a secondary, yet related reason for the increase in the use of reclaimed timber, in that certain desired species of timber have become rare or expensive to obtain, so reclaimed hardwoods such as mahogany and teak may occupy a high value niche. This may even be the case for softwood species such as pitch pine from nineteenth century buildings, where growth rate and strength properties are highly appreciated.

Currently, the estimated trade in salvaged timber represents a relatively small market share, which has decreased by 21 % between 1997 and 2007 (CRWP 2009). This may partially reflect an increased use of the internet leading to a reduction in volume passing through the traditional salvage merchants and architectural traders. There may also be a trend for high value products to pass through community recycling projects and return lower prices to ensure rapid sales, as opposed to the traditional longer stock holding periods associated with salvage trade which led to higher product price from the renovation customer set.

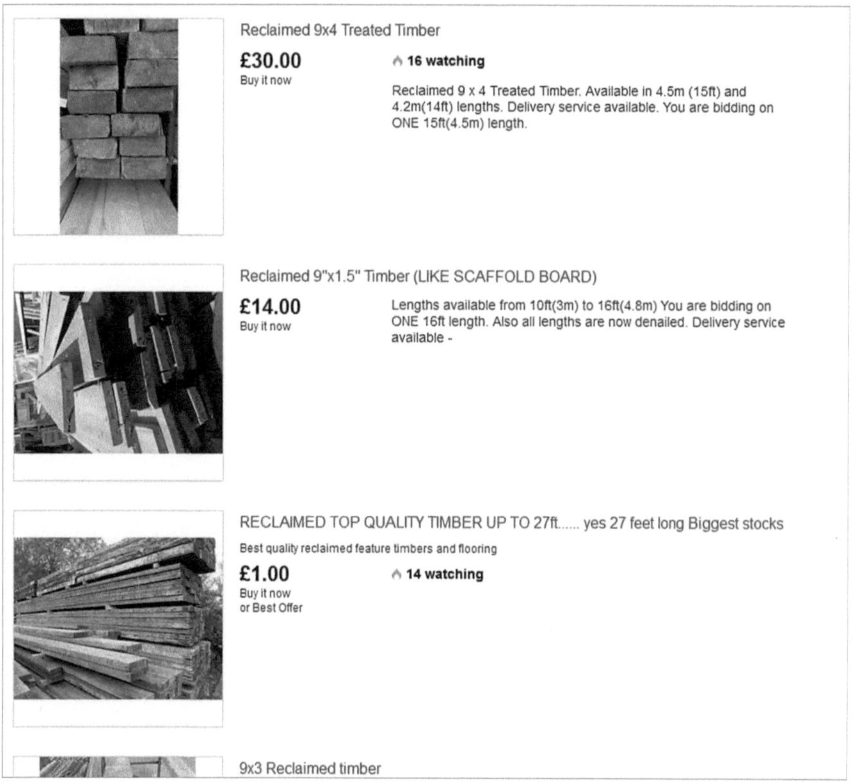

Fig. 16 Online sales of reclaimed timber are increasingly connecting customers with those holding stocks from renovation and demolition works

The use of reclaimed timber is not without its problems, for example the history of the timber may well be unknown. Essentially, whether the timber has been treated with preservatives—some traditional now withdrawn preservatives require special care and specific personal protective equipment when sawing or planing. Likewise lead-based paints should be avoided for certain re-use applications if children may come into contact with the product. In addition there is the question of whether there are larger contaminants, such as metal staples, embedded within the timber which could cause problems when re-sawing or installing a beam to a new location. However, in spite of the inherent issues and potential contamination problems associated with the use of reclaimed wood, the industry has been shown to be economically sound. Shi et al. (2001) showed that the removal and reprocessing of creosote timber pilling from a US naval base saved approximately $32/tonne in addition to the landfill tax avoidance.

In terms of environmental performance, such reuse extends the lifetime of the original material (the timber) into a second service life, thereby avoiding the consumption of an equivalent quantity of virgin material and, importantly, sequestering the carbon contained within it for an extended period. This 'double benefit' can be realised as long as reclamation, sorting and reprocessing procedures are environmentally light. Of these the reprocessing stage is perhaps the key, since the removal of chemical contaminants, such as preservatives, resins, etc., is likely to be the most environmentally intensive of these steps, if chemical or enzymatic cleaning is undertaken. Many aspects of the process optimisation, and the environmental impact of different strategies in sorting segregating and cleaning the reclaimed wood remain to be investigated, in working towards successful and environmentally responsible onward use of the timber.

5.2 *Recycling Back into Construction Materials*

The dominant mode of recycling demolition wood back into the building industry has already been discussed in Sect. 2, this being the manufacture of particleboard from waste wood. It has been shown that the composites industry, whilst embracing 'clean' waste wood, does not use demolition wood in their products due to the potential contamination and the effects this will have on the final products. In recent years, researchers have developed methods for detection of treated wood in the wood waste stream, and for the upgrading of treated wood to clean wood, however, many of these have yet to become a reality due to economic constraints (Helsen and Van den Bulck 2005).

5.2.1 Cleaning Wood for Panel Products Manufacture

The wood panel industry have embraced the use of recycled wood and currently within the UK, all particleboard is manufactured from recycled timber. However this timber has to be graded as clean (from sources such as joinery waste and pallet wood) and demolition wood does not fall within this category. Researchers have addressed methods for sampling and quantification of contaminant levels and their influence on board properties (Irle and Ormondroyd 2002; Irle et al. 2003; Fru et al. 2003), and in recent years have been assessing technologies for better sorting systems, and fractionation of the waste stream in an effort to utilise more waste wood (Poon et al. 2001; Yasuda et al. 2006; Kumbhar et al. 2013).

In addition, systems such as chemical treatments, biological treatments and liquefaction have been tested, to remove preservative treatment chemicals from the wood. In some cases, the technology has focused on reclamation of the copper, chromium and arsenic and conversion to a form suitable for re-use as a wood treatment chemical (Gezer and Cooper 2009; Janin et al. 2009), others have investigated the production of cleaned chips for future re-use (Kartal and Clausen 2001; Coudert et al. 2015). The pilot scale work by Coudert et al. (2013) produced wood chips with a significantly reduced copper content from CCA, ACQ, MCQ and copper azole treated woods. Up to 97.5 % of the arsenic, 87.9 % of the chromium and 96.1 % of the copper present in CCA treated wood was removed by the three step leaching process. The particles were sufficiently clean for compost production, but only suitable for use in particleboards if mixed with clean wood chip at 10–50 % of the chip feedstock (US regulations).

Work with bacteria such as *Lactobacillus* and *Streptococcus* shows some potential (Chang et al. 2012) however may require several days to achieve metal removal. The laboratory study used milled wood powder so may generate material better suited to WPC production than particleboard. Milling was shown to be unnecessary for bacterial fermentation by *Bacillus licheniformis* when oxalic acid was used as a pre-treatment (Clausen and Smith 1998). Electrokinetic processes have been investigated at the bench scale, for example Sarahney et al. (2005) demonstrated the removal of 74 % of chromium, 97 % of the copper and 88 % of the arsenic from CCA treated timber. The electrokinetic process was enhanced with a oxalic acid-EDTA solvent mixture. Partial liquefaction is a novel system which may offer particle cleaning for recycling into panel products (Medved et al. 2015). Whilst there has been success at a small scale level for many different approaches, the move from lab to industry has not yet taken place.

5.2.2 Recycling Fibreboard Panels for MDF Manufacture

Another aspect of recycling from the demolition site to the panels industry is the recycling of panel products from demolition back into the panels industry. Whilst this has been the subject of research for some years (Kearley and Goroyias 2004) one company is pioneering the use of recovered MDF fibres back into the MDF industry, along with others. MDF recovery have patented technologies for the separation of fibres from boards and have proven their use in the remanufacture of MDF boards with similar board properties and emission profiles to those of virgin boards (Bartlett 2015).

5.2.3 Wood-Plastic Composites

WPCs, similarly to MDF, traditionally use virgin timber to ensure a consistency in the wood particle feedstock used in the manufacturing process. This is partially for minimisation of off-gasing of volatiles such as terpenes from softwood timbers, and partially to ensure efficient particle size production, or simply due to co-location with existing wood processing industries which are generating sawdust or planer shavings as waste.

The WPC product is a blend of thermoplastic matrix with wood particles or wood fibres, depending on manufacturer and technology employed. Polymers may include polyethylene, polypropylene, PVC or polystyrene, or may increasingly be biopolymers, such as polylactic acid (PLA) or thermoplastic starch. Each has a different target market and attributes, relating to the polymer characteristics, however, a commodity market surrounding extruded decking planks and cladding sections has become well established (Spear et al. 2015) with a production of 172 thousand tonnes per annum in Europe alone. There is scope to recycle demolition wood into the WPC product, especially where relatively large quantities of uniform timbers such as studwork or joists are available from a site.

When recycled wood is used, the presence of large objects (nails, fixings etc.) presents a challenge to the milling equipment, or the extruder if small fragments pass into the wood particle feedstock. The other contaminants such as paint, varnish or treatment products and glues may contribute additional considerations such as degradation in the high temperature of the process, or a reduction of compatibility between the particles and the polymer matrix.

However, recent research has evaluated the use of waste wood in the manufacture of WPC products. Nourbakhsh et al. (2010) evaluated the use of wood flour derived from sawdust waste and waste polypropylene in manufacturing WPCs. The results of this work indicated that the use of the waste materials showed promise as an effective alternative raw material to virgin wood. In fact, many WPC wood feedstocks are derived from sawdust residues collected from sawmills, ensuring uniformity of species, particle size and moisture content (Spear et al. 2015).

Chaharmahali et al. (2010) manufactured WPC panels using the hot press method and with sawdust from MDF and particleboard manufacture plants. They added the wood at different percentages and compared mechanical properties to those of virgin MDF and particleboards. Water soak tests showed lower weight gains in the WPC material, as expected, while MOR was poorer than for MDF. However, comparison with generic data for WPC materials (Spear et al. 2015) indicates that the modulus of elasticity (2.5–3.0 GPa), Modulus of rupture (25–35 MPa) and unnotched impact strength (400–550 J/m) values obtained at 60 % fibre loading were within the range expected for uncompatibilised WPCs. The MDF sawdust WPC product formed by Chaharmahali et al. performed better than the particleboard sawdust, which was attributed to the fibrous quality of the MDF sawdust. Similar results may be expected if post-consumer or demolition waste MDF were segregated and recycled into a WPC product. Recycled newsprint and other paper fibres have been used to form WPC materials with improved properties, relating to the fibrous rather than particulate character of the filler (English et al. 1997).

Elsewhere Xu et al. (2008) have looked at the environmental performance of wood fibre reinforced polypropylene composites and compared this with pure polypropylene. They tested composites of varying fibre content (10, 30 and 50 % by mass) in order to establish a functional unit based on tensile strength and then performed LCA to compare the two. Although their study was based on fibre produced from logging residues, and is therefore not directly comparable to that recovered (and reprocessed) from demolition sites, the composite did show favourable results in all impact categories compared to 100 % polymer.

Whilst the research on the use of waste demolition wood in the manufacture of WPCs is limited, and indeed eclipsed by the research in the use of recycled plastics, WPC are a promising product for the use of demolition wood. Once again, the recurring theme of cleaning the timber from contaminants must be taken into account.

5.2.4 Wood Cement Composites

Wood cement composites have been investigated for over a century and industrialisation began in the 1930s (Frybort et al. 2008). The panels have been used for a variety of applications including insulation (Wolfe and Gjinolli 1996) acoustic insulation (Frybort et al. 2008) and for structural applications (Papadopoulos et al. 2006; Miyatake et al. 2000). They are frequently selected for their resistance to fire (Soroushian et al. 2013; Spear 2015) and to termites (Eusebio 2003; Wolfe and Gjinolli 1996).

Many different types of cement-bonded composites have been produced, including wood wool cement boards, cement-bonded particle boards (Soroushian et al. 2013), cement bonded oriented strand boards (Ntalos and Papdopoulos 2006) and cement-bonded composite beams (Bejo et al. 2005; Datye and Gore 1998). In an early study on possible re-use options for CCA-treated wood in Florida, cement-bonded particleboards and wood cement composites were identified as having potential for several reasons. The wood may reduce the density of the pure cement, giving better insulating properties, while the cement may contribute to stabilisation of the metals within the wood, minimising potential for leaching (Solo-Gabriele and Townsend 1999).

Some research has been undertaken to assess the properties of cement-bonded wood composites manufactured with demolition wood (Zhou and Kamdem 2002; Wolfe and Gjinolli 1996; Qi et al. 2006). Leaching of copper and arsenic was greatly reduced for Portland cement composites including CCA-treated wood particles, however, chromium remained leachable (Huang and Cooper 2000). Schmidt et al. (1994) indicated that CCA treated wood had greater compatibility with the cement than untreated wood, with greater resistance to fibre pull out.

The species of wood used in cement composites can alter cure rate, with species which are high in sugars and tannins offering the greatest challenge (Weatherwax and Tarkow 1964; Wei et al. 2000; Karade 2010). It is clear that wood cement composites have the potential to utilise demolition wood, but it has been shown that the composition of the demolition wood can have an effect on the curing of the concrete. Thus, the manufacturing system will have to be tuned to wood type and therefore composition of the substrate will ideally be known (Qi et al. 2006; Wei et al. 2000). The product is also relatively expensive to produce per square metre, compared to other wood-based panels, due to the slow curing rate of the cement. Efforts to increase curing rate and process efficiency have been investigated, as reviewed by Frybort et al. (2008) and Karade (2010).

Recycled wood from demolition and construction is likely to be best suited to the production of particles for cement-bonded particleboard, as this reduction technology is better developed and more suited to the variable dimension input material and removal of contaminants such as nails and screws. Excelsior (wood wool) for wood wool cement composites requires large blocks for manufacture, and strands for cement-bonded OSB require logs of relatively uniform dimensions for efficient production. However, Qi et al. (2006) showed that MDF can be recycled into wood cement composites with good results.

LCA data does currently exist for cement-bonded particle board, based on boards with 20 % w/w wood and production data from 2000 (Werner et al. 2007b). This could be developed further to account for the incorporation of reclaimed, sorted demolition timber into the mix.

6 Demolition Waste Wood to Fuel

It has often been noted that the use of wood as a source of energy would help the world reduce its CO_2 emissions and help the energy sector hit the targets set out in the Kyoto protocol (Skodras et al. 2004). However, as we have already noted the simple combustion of some waste woods can lead to environmental risks due to their chemical nature and due to the presence of organic additives which can result in the release of toxic compounds, such as dioxins and furans (Addink and Olie 1995; Blumenstock et al. 1998). Other researchers have also commented on the value of the waste wood commodity and that a route of materials cascading would derive the maximum useful life from the wood fibre source and that its use as fuel should be the last stage in the cascade (Hill 2011).

However, new methods of utilisation of waste wood have been the subject of research. These have included co-utilisation of the demolition wood with other feedstocks, gasification of the demolition wood and torrefaction. Within this section the new technologies will briefly be explored and their applicability to the use of demolition wood be assessed.

6.1 Direct Combustion and Co-firing of Demolition Wood

The direct combustion of demolition wood to produce heat and energy is generally seen as the final option in the waste hierarchy (alongside land fill), however, it has been noted that the replacement of fossil fuels with wood would reduce net atmospheric CO_2 by 90 % (De Jong et al. 2003; Hagedorn et al. 2003). Demolition wood is generally classed as low grade fuel mainly due to the contamination, whether macro, micro or chemical. The contamination of the timber has been shown to cause issues in both direct combustion and in co-firing situations. Yorulmaz and Atimay (2009) showed that in TGA analysis wood-based composites showed a lower rate of weight loss and a higher char yield than with the control samples. Amand et al. (2006) concluded that there was an increased ash builds up on the heat transfer tubes within wood burning boilers when demolition wood was used, whilst Elled et al. (2008) concluded the same fate for zinc deposits.

Helsen and Van den Bulck (2005) studied model systems of CCA treated wood to observe the fate of combustion products of the arsenic, copper and chromium.

They concluded that co-incineration was the most appropriate technology in the short term, due to the relatively low investment costs, relative ease of complying with emissions legislation due to dilution by the co-fired material, and economies of scale. In the longer term, low temperature pyrolysis, or high temperature gasification systems were preferable options for disposal of CCA treated wood. The slow pyrolysis offers greater control of arsenic volatilisation, avoidance of dioxin formation and metal recovery. Gasification offers no dioxin formation, high energy efficiency and potential to recover metals and fuels.

Whilst the issues with the emissions profile of demolition wood still remains, new technology in both the combustion chamber and in the scrubbing of the flue gases continues to improve. Continued process innovations will lead to cleaner combustion of demolition wood waste. The process controls are likely to lead to environmentally benign energy generation from these wood residues.

6.2 *Torrefaction*

Torrefaction is the up-grading of biomass by a thermal pre-treatment to make a more uniform and attractive biofuel (van de Selt et al. 2011). The main purpose of torrefaction is to remove the oxygen from the biomass and increasing the C/O ratio. This produces a fuel which has properties much more similar to coal for handling in an industrial or domestic context. Torrefaction is carried out at temperatures between 200 and 300 °C, atmospheric pressure and in the absence of oxygen.

Torrefaction has been utilised for the upgrade of many timber species such as birch, pine (Pach et al. 2002) bamboo (Indian Institute of Science 2006), oak (Pierre et al. 2011) and wood briquettes (Felfli et al. 2005). Researchers have shown that demolition waste wood can be torrefied to upgrade to produce a material with properties similar to those of torrefied virgin timber (Li et al. 2012; Medic et al. 2012; Tumuluru et al. 2011). Where such wood is contaminated to begin with, processes will be needed to ensure that the resulting fuel is used under appropriate conditions.

In environmental terms, Adams et al. (2015) have compared LCA output for torrefied and non-torrefied Scots Pine pellets and found that the torrefied pellets performed better when considered on a ‘per MJ energy delivered’ basis (for both GHG and fossil energy profiles). Their results were highly sensitive to assumptions made about the drying stage of the process (specifically, the energy requirement for drying) and the proportion of heat supplied from re-circulated torrefaction gas. These are therefore the key stages for focus in terms of further development of this process. The positive results in the GHG and fossil energy profiles should be considered alongside the higher land use requirement for the torrefied pellets (on account of the mass lost during production), in order to give the fullest picture of this technology.

6.3 Gasification of Demolition Wood

Gasification is an established technology (Rezaiyan and Cheremisinoff 2005; Quaak et al. 1999) with the ability to produce synthetic gas from carbon rich feedstocks. Gasification has recently received a resurgence in attention and the technology has evolved to minimise the carbon footprint of the operation, with some research assessing the applicability of other construction waste (namely concrete and mixed waste) as an absorbent of CO_2 which has led to the sustainable production of hydrogen (Moghtaderi et al. 2012).

The gasification process is done by the reaction of the carbon source with an oxidising agent (e.g. steam, oxygen and/or air) in a reaction at elevated temperatures (500–1500 °C) and pressures (up to 33 atms). When processed at a temperature of 1100–1500 °C all organic compounds are cracked, which eliminates the risk of polychlorinated dibenzo-*p*-dioxins and polychlorinated dibenzo-*p*-furans being formed by combustion in the presence of copper (Helsen and Van den Bulck 2005), this gives gasification a significant advantage over traditional combustion of energy operations. Gas cleaning equipment is still required, however for a smaller volume of gas. High temperature gasification operations may also increase efficiency of arsenic removal due to formation of metallic arsenic in place of arsenic trioxide.

The gasification of demolition wood is still relatively rare, however, with significant growth potential. One barrier is the heterogeneous state of the wood as it enters the system, however, demolition waste has been used as a feedstock in the Netherlands. The Amer demolition wood gasification project (Willeboer 1998) concluded that the use of an existing modern coal fired power station the gasification of (low quality) demolition wood with a high conversion efficiency and an 'outstanding' environmental performance could be achieved at a relatively low capital cost. Faaij et al. (1997) modelled the use of several biomass wastes for gasification, including demolition waste. The research showed that demolition wood was currently available at a negative cost (i.e. you can charge to take it away) and that it is a drier fuel source than green material. The models used within the paper also showed that demolition waste exhibited a good conversion efficiency at 40.0 %, which is comparable to the same model run for clean wood (40.3 %). It is important to note that the calculated cost per KWh for electricity generated using demolition wood gasification could be as low as -€4.8, depending on the capital cost of the plant. This potential net gain is due to the negative cost of the fuel mentioned above.

It has been shown that the gasification of demolition wood is a viable option, as long as the wood is clean of macro-contamination before it can be used in these systems.

7 Future Opportunities and Challenges

7.1 The Biorefinery Concept

The biorefinery concept is one that integrates the conversion of a single biomass source into multiple products maximising the yield and therefore the value derived from the feedstock. A biorefinery is analogous to an oil refinery, which produces multiple fuels and chemicals from a single, crude oil, feedstock. In the same manner, the biorefinery also produces a range of fractions or products based on the single biomass feedstock (Bennet and Pearson 2009; Charlton et al. 2009).

Currently the 'whole crop biorefinery', the 'green biorefinery' and the 'lignocellulosic-feedstock biorefinery' systems are favoured in research and development (Kamm and Kamm 2004). While the whole crop biorefinery is well suited to segregation and use of straw, seed and chaff from crops, and the green biorefinery is designed to recover sugar-rich sap, dyes and other components from crops such as grasses, the lignocellulosic biorefinery concept could be applied to reclaimed wood in addition to the woody materials currently discussed. More recently, a pyrolysis oil-based biorefinery concept has been discussed in which the main chemical groups (acetic acid, glycolaldehydes and aldols) are recovered from pyrolysis oil of forest residues or similar materials (Vitasari et al. 2015). The pyrolysis oil system may have application for reclaimed wood. Ferreira et al. (2015) demonstrated the generation of char, bio-oil and fuel gas by pyrolysis of MDF residues from furniture production. This presents an energy recovery aspect and some platform chemical production, whereas the three main biorefinery types focus on generation of a suite of products with energy as the final co-product.

Whilst there is currently a great interest in the biorefinery concept and indeed there are several fuel-focused biorefineries operating today (for example Blue Marble Energy, Himark BioGas and Chemrec's integration of a black liquor gasification plant and biomethanol plant with a pulp mill) the concept is yet to be fully realised.

In the development of the biorefinery concept there has been some interest in the development of waste centred biorefineries, although these are generally focused on the use of food waste and primary production waste (Fava et al. 2015). However, there is an opportunity to develop a biorefinery that incorporates technologies to effectively deal with the potential contamination that arises with the use of demolition wood and indeed other waste wood resources. One example may be the use of certain fungal strains to decontaminate chlorinated phenols, dibenzo-*p*-dioxins and furans within waste wood or soil contaminated by now withdrawn wood treatment chemicals (Steffen et al. 2002; Valentin et al. 2013). This would potentially create a product line based on the contamination. Likewise, work with oxalate producing fungi such as the basidiomycetes and *Aspergillus niger* has demonstrated capability to remove metal ions typical in the CCA-type wood treatments (Kartal et al. 2004a; Kim et al. 2009). The copper removal system has been evaluated for use with recovered wood (Kartal et al. 2004b). No work has yet exploited this bioremediation capability of the fungi specifically for application within a biorefinery, however, the concept would be the same. The fungally pretreated wood could be passed through a sequence of mechanical, microbial or enzymatic steps, and extraction or separation stages produce different products. It is expected that all technologies would use environmentally sustainable means, and the range of bio-derived products entering different markets would ensure the economical sustainability of the whole operation.

7.2 Nano-Cellulose Production

Nano-cellulose is a material derived from wood fibres, with exceptional strength characteristics, on a par with Kevlar. However, unlike fossil fuel and Kevlar derived fibres, nano-cellulose is renewable. The manufacture of nano-cellulose has, in recent years, been developed from an energy intensive process ($\sim$30,000 kWh/tonne) to around 500 kWh/tonne a reduction of 98 % (Klemm 2011). There are many applications for nano-cellulose including food thickeners (Strom et al. 2013) nano-composites (Siro and Plackett 2010; Eichhorn et al. 2010), cosmetics (Klemm et al. 2006), applications in the electronics sector (Torvinen et al. 2012) and in insulation (Wicklein et al. 2015).

Recent research into the production of nano-cellulose from 'waste' resources has focused on the use of agri-waste (Rosa et al. 2010; Mandal and Chakrabarty 2011; Xiong et al. 2012) however, there is potential for waste (and indeed demolition) timber to be used for this purpose. However, it is likely that this would have to be an integrated part of a biorefinery concept as described above as the cleaning of the waste source will be paramount in a successful process.

7.3 Challenges

One significant challenge confronting the recovery, recycling and use of wood waste is the intertwined issues of diversity and dispersal. The wide range of materials, of a wide range of ages, all of which may contain many different generations of treatment agents, paints or other products, become mixed at the construction or demolition site. These are further combined and mixed on leaving the site for on-processing at a recycling plant. In many recycling facilities the first operation after magnetic removal of large metal contaminants is crushing or grinding to smaller particulate, which homogenises the size of particles but further mixes the feedstocks. At this point in time the chips are unrecognisable as having been softwood, hardwood, treated fencepost, painted joinery or chipboard in their previous life.

One simple step is the introduction of segregation bins on arrival at the reprocessing plant, to separate particleboard from softwood, possibly also segregating other classes such as MDF or hardwoods in additional bins. Space, cost and staff training are key determinants in this scenario. However, segregation using spectrometric systems to detect CCA-treated timber could be introduced within this framework, to generate distinct product classes. Labelling could assist this process, if timber were marked as treated in an easily recognisable manner. This may be especially useful for products in lower hazard classes such as window frames, where diffusion treatment by borates, or organic fungicides in light organic solvent systems may have been used. These treatments are less visible than the characteristic copper colouration of CCA systems (although studies have shown that even CCA cannot be reliably visually recognised in sorting reclaimed wood).

Many structural timber elements used within construction are marked with a grade stamp—for example C16 or C24 whitewood lumber is clearly identifiable at point of sale. Timber merchants have greater responsibility under the EU Timber Regulations (EUTR) to perform due diligence on any products sold, in particular this is to control and prevent the sale of illegally harvested timber in the EU. As a result, merchants have placed an increasing emphasis on maintaining records for all batches of timber, so that full information can be provided for their products. The various EN standard grade stamps for timber and panel products, which are widespread across Europe, have also been better communicated between suppliers and customers. So in future a greater degree of recognition of product, and identification of timber which was sold kiln dried, graded, or with preservative treatments, will become more straightforward where large pieces of timber are salvaged during demolition. Current problems relate partly to large-scale demolition by machine, where timber is removed from rubble after the building has been knocked down—thus structural timbers are mixed with other fittings from internal furniture, window and door frames, etc. Costs for sorting waste wood relate largely to the labour involved (Jacobi et al. 2007), so greater automation at the recycling facility is ultimately a more economically viable system.

In a Swedish study on the waste handling chain for recovered wood waste, some good examples of waste segregation near the source were identified (Krook et al. 2008). However, the study also revealed that in most cases the legislation alone was an insufficient driver for change and best practice adoption in the other processing plants. Market forces encourage those handling the waste to rely on an end-of-pipe solution rather than taking simple steps to improve segregation at an early stage, which could lead to higher volumes of higher value material for future re-use. However, WRAP have produced clear simple guidance for those handling recovered wood, to assist in sorting at source (Fig. 17, WRAP 2015). The National Community Recycling Project has generated social enterprises across England, and in some parts of Wales and Scotland, to engage in sorting the highly mixed wastes derived from builders skips (Community Wood Recycling 2014). This has shown a steady increase in total volume collected, and in proportion recycled or reused, for example in 2014 40 % of the total wood collected was reused.

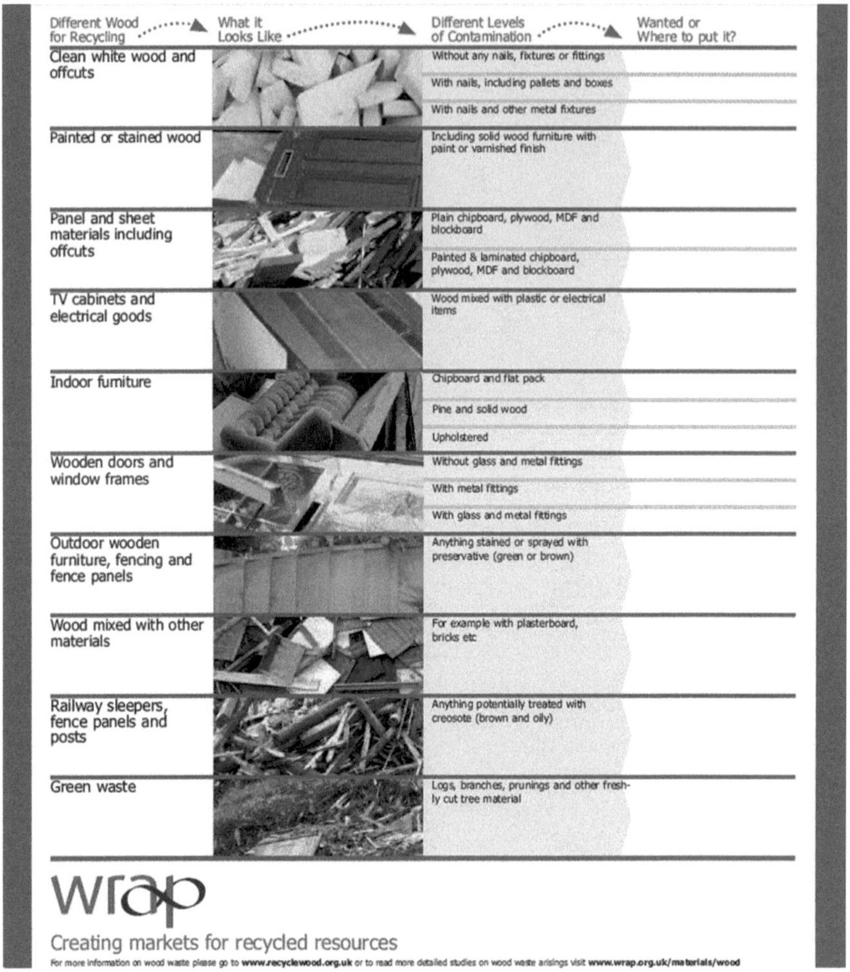

Fig. 17 Guidelines produced by WRAP for wood recyclers in the UK to aid classification and sorting of waste wood

The segregation process becomes significant if a plant were to seriously invest in cleaning technologies for recycling wood chips into particleboard, as there is little to gain in washing a product which is 80 % clean timber with expensive and time consuming solvent sequences. Greater efficiency could be achieved by demonstrating that the clean 80 % had been removed and chipped prior to cleaning, with the 20 % of contaminated wood then chipped and cleaned to an approved extent for re-sale into the target market. Other dedicated recycling schemes such as MDF Recovery's process for recycling MDF fibre also rely on segregation of the target fraction from the wood gathered at civic amenity sites, or the demolition wood fraction.

7.4 Summary

The potential options for the reuse and recycling of demolition wood in future are vast and range from the use of the timber as reclaimed wood through to the use as a feedstock for a biorefinery and for the manufacture of nano-cellulose. The options presented above are in various stages of research, some still in the laboratory whilst some are being evaluated for their economic feasibility. However, many of the options presented above relay on the demolition wood been cleaned and free of contamination. Until an economical and viable method for the cleaning of demolition wood is found the wood will continue to go to land fill or low grade energy recovery.

8 Conclusions

This chapter has demonstrated that the waste wood from C&D projects can be utilised in a profitable and environmentally friendly way. However, this is not without its challenges. The lack of knowledge of the history of the timber being removed from building sites leads to conservative use of the timber and in some cases an extreme reluctance to process and utilise the resource.

However, legislation in its vast array of forms has both driven and hindered the reuse and recycling of demolition timber. On the one hand legislation has encouraged the use of demolition timber, whether this is by incentives, such as renewable obligation certificates or by disincentives, such as landfill taxes. Whilst at the same time the use has been hindered by legislative barriers such as categorisation of demolition waste of unknown origin as hazardous waste.

The conflicting legislation has led to recent developments in the screening and sorting of timber from demolition sites and this had seen to an increase in the recycling and reuse of the timber. A continuation of the development of the techniques along with a cost-effective commercialisation programme will lead to a greater utilisation of demolition timber.

In addition, the improvement in the sorting techniques the recycling pathways and options have become greater. Developments in the use of clean recycled timber into panel products, WPCs and cement wood composites has led to a market for clean timber beyond the traditional reclamation and reuse and the energy recovery scenarios. The development of processes for recycling wood panel products from demolition sites, whilst still in its infancy, has been proven and is currently in the commercialisation phase, which, once this has occurred, will lead to an increase in the removal of MDF from the waste stream.

There is a wealth of published LCA data relating to wood and wood products as used in the construction sector, though this is most prevalent for products in the first service life of the materials cascade. Further, research assessing the whole-lifetime impacts of wood material as it passes through progressive recycling and reprocessing steps would be welcome. Similarly, further harmonisation of the various standards governing LCA methodology in general (and in relation to wood and construction, in particular) can only be a good thing in terms of encouraging its wider uptake. At the EU level, it is the stated aim of the PEF programme to achieve this and this will draw to its conclusion in the coming 2 years.

Future developments for the use of demolition wood that can utilise timber with an unknown history will vastly change the markets for the resource. The reduction of the timber to its simplest components that can then be cleaned within the process, for example, through an integrated refinery approach, will lead not only to more environmentally friendly and economically viable system but ultimately to a no waste scenario.

References

Adalberth K (2000) Energy use and environmental impact of new residential buildings, Report TVBH-1012, Department of Building Physics, Lund Institute of Technology, Lund

Adams PWR, Shirley JEJ, McManus MC (2015) Comparative cradle-to-gate life cycle assessment of wood pellet production with torrefaction. Appl Ecol 138:367–380

Addink M, Olie K (1995) Mechanismss of formation and destruction of polychlorinated dibenzo–dioxins and dibenzo-furans in heterogeneous systems. Environ Sci Technol 29(6):1425–1434

Aktas CB, Bilec MM (2012) Impact of lifetime on US residential building LCA results. Int J Life Cycle Assess 17:337–349

Amand L-E, Leckner B, Eskilsson D, Tullin C (2006) Ash deposition on heat transfer tubes during combustion of demolition wood. Energy Fuels 20:1001–1007

Anon (2008) The rise and rise of timber frame. Timber in construction. http://www.timberinconstruction.co.uk/features/rise-and-rise-timber-frame. Accessed 18 Nov 2015

Asif M, Davidson A, Muneer T (2002) Life cycle of window materials—a comparative assessment. Millennium Fellow School of Engineering, Napier University, Edinburgh. http://www.researchgate.net/publication/237446892. Accessed 25 Aug 2015

Athena Sustainable Materials Institute (2012) A cradle-to-gate life cycle assessment of Canadian glulam manufacture: an update. http://www.athenasmi.org/wp-content/uploads/2013/10/CtoG-LCA-of-Canadian-Glulam-Manufacture-Update.pdf. Assessed 25 Aug 2015

Bartlett C (2015) MDF: closed loop recycling—enhancing supply chain value. In: Spear M et al (eds) Proceedings of the international panel products symposium 2015, Llandudno, Wales, pp 83–88, 7–8 Oct 2015

Bejo L, Takats P, Vass N (2005) Development of cement bonded composite beams. Acta Silv Lign Hung 1:111–119

Bennet SJ, Pearson PJG (2009) From petrochemical complexes to biorefineries? The past and prospective co-evolution of liquid fuels and chemicals production in the UK. Chem Eng Res Des 87(9):1120–1139

Blumenstock M, Zimmermann R, Schramm KW, Kettrup A (1998) Low Chlorinated benzenes, dibenzo-*p*-dioxins and dibenzofurans as surrogate compounds for the PCDD/F emissions in the flue and stack gas of hazardous waste incinerator. Organohalogen Compd 36:59–63

Borjesson P, Gustavsson L (2000) Greenhouse gas balances in building construction: wood versus concrete from life-cycle and forest land-use perspectives. Energy Policy 28:575–588

Bovea MD, Vidal R (2004) Increasing product value by intergrating environmental impact, costs and costomer valuation. Resources, Conservationanf Recycling 41(2):133–145

BRE (2004) Wood waste management—UK updated (cited in WRAP 2009 report)

Bribián IZ, Capilla AV, Usón AA (2011) Life cycle assessment of building materials: comparative analysis of energy and environmental impacts and evaluation of the eco-efficiency improvement potential. Build Environ 46:1133–1140

BSI (1993) BS EN 71-5:1993 safety of toys. Chemical toys (sets) other than experimental sets. British Standards Institute, London

BSI (2004) PAS 104 wood recycling in the panel board manufacturing industry. British Standards Institute, London

BSI (2011) PAS 2050:2011, specification for the assessment of the life cycle greenhouse gas emissions of goods and services. British Standards Institution, London

BSI (2012) PAS 111:2012 specification for the requirements and test methods for processing waste wood. British Standards Institute, London

BSI (2013) BS EN 15804:2012+a1:2013 sustainability of construction works—environmental product declarations—core rules for the product category of construction products. British Standards Institute, London

Chaharmahali M, Mirbagheri J, Tajvidi M, Najafi SK, Mirbagheri Y (2010) Mechanical and physical properties of wood-plastic composite panels. J Reinf Plast Compos 29(2):310–319

Chang YC, Choi DB, Kikuchi S (2012) Enhanced extraction of heavy metals in the two-step process with the mixed culture of Lactobacillus bulgaricus and Streptococcus thermophiles. Bioresour. technol 103(1):477–480

Charlton A, Elias R, Fish S, Fowler P, Gallagher J (2009) The biorefining opportunities in Wales: understanding the scope for building a sustainable biorenewable economy using plant biomass. Chem Eng Res Des 87(9):1147–1161

Clausen CA, Smith RL (1998) Removal of CCA from treated wood by oxalic acid extraction, steam explosion and bacterial fermentation. J Ind Microbiol Biotechnol 20(3–4):251–257

Community Wood Recycling (2014) Waste wood in the UK. http://www.community-woodrecycling.org.uk/about-us/our-impact/. Accessed 13 Nov 2015

Coudert L, Blais J-F, Mercier G, Cooper P, Gastonguay L, Morris P, Janin A, Reynier N (2013) Pilot-scale investigation of the robustness and efficiency of a copper-based treated wood wastes recycling process. J Hazard Mater 261:277–285

Couret L, Irle M, Belloncle C, Cathala B (2015) Extracting high-value products from waste MDF. In: Spear M et al (eds) Proceedings of the international panel products symposium 2015, Llandudno Wales, pp 89–96, 7–8 Oct 2015

CRWP (2009) Overview of demolition waste in the UK. Construction resource and waste platform. (CRWP, www.crwplatform.co.uk now defunct, but data available from www.wrap.org.uk/content/crwp-publications), http://www.wrap.org.uk/sites/files/wrap/CRWP-Demolition-Report-2009.pdf

Datye KR, Gore VN (1998) Innovative designs of wood-bamboo-brick composites In: Proceedings of the 5th world conference on timber engineering Montreux, Switzerland, pp 526–533

Davies I (2009) Sustainable construction timber: sourcing and specifying local timber. Forestry Commission Scotland, Edinburgh, 30 p

De Jong W, Pirone A, Wojtowicz MA (2003) Pyrolysis of *Miscanthus giganteus* and wood pellets: TG-FTIR analysis and reaction kinetics. Fuel 82:1139–1147

DEFRA (2010) Environmental permitting guidance, the waste incineration directive: for the environmental permitting (England and Wales) regulations 2010. PDF document 96 p. https://www.gov.uk/government/uploads/system/uploads/attachment_data/file/201215/pb-13570-wid-guidance-201003.pdf. Accessed 15 Nov 2015

DEFRA (2015) Greenhouse gas conversion factor repository. http://www.ukconversionfactorscarbonsmart.co.uk/. Accessed 25 Aug 2015

DTI & DEFRA (2007) UK biomass strategy (cited in WRAP 2009 report)

Eichhorn SJ, Dufresne A, Aranguran M, Marcovich NE, Capadona JR, Rowan SJ, Weder C, Thielemans W, Roman M, Rennecker S, Gindl W, Veigel Keckes J, Yano H, Abe K, Nogi M, Nakagaito AN, Mangalam A, Simonson J, Benight AS, Bismark A, Berglund LA, Peijs T (2010) Review: current international research into cellulose nanofibres and nanocomposites. J Mater Sci 45:1–33

Elled A-L, Amand L-E, Eskilsson D (2008) Fate of Zinc during combustion of demolition wood in a fluidized bed boiler. Energy Fuels 22:1519–1526

English B, Clemons C, Stark N, Schneider JP (1997) Wastewood-derived fillers for plastics. In: Proceedings of the fourth international conference on woodfiber plastic composites. Madison, WI, pp 309–324, 12–14 May 1997

EPF (2002) EPF Standard for delivery conditions for recycled wood. Panel guide 2002. http://www.europanels.org/upload/EPF-Standard-for-recycled-wood-use.pdf. Accessed 7 Dec 2015

ERM (2006) Carbon balances and energy impacts of the management of UK wastes (cited in WRAP 2009 report)

EU JRC (2012) Product environmental (PEF) guide. http://ec.europa.eu/environment/archives/eussd/pdf/footprint/PEF%20methodology%20final%20draft.pdf. Accessed 26 Aug 2015

European Commission (1999) Landfill Directive (1999/31/EC)

European Commission (2000) Waste Incineration Directive, EC Directive (2000/76/EC)

European Commission (2008) Waste Framework directive (2008/98/EC)

European Commission (2009) Renewable Energy Directive (2009/29/EC)

European Commission (2010a) International reference life cycle data system (ILCD) handbook—general guide for life cycle assessment—detailed guidance, 1st edn. Publications Office of the European Union, Luxembourg

European Commission (2010b) Industrial Emissions Directive (2010/75/EU) Dec 2010

Eusebio DA (2003) Cement bonded board: todays alternative In: Technical forum in celebration of the PCIERD 21st anniversary, Pasig City, The Philippines, pp 23–45

Fava F, Totaro G, Diels L, Reis M, Duarte J, Beserra Carioca O, Poggi-Varaldo HM, Ferreira BS (2015) Biowaste biorefinery in Europe: opportunities and research & development needs. New Biotechnol 32(1):100–108

FAO (2010) Forest Products Statistics 2005-2009, Timber Bulletin ECE/TIM/BULL/63/2 – provisional http://timber.unece.org/index.php?id=84 Accessed 15/3/11

Felfli FF, Suarez CA, Luengo JA, Beaton P (2005) Wood briquette torrefaction. Energy Sustain Dev 9(3):19–22

Ferreira SD, Altafini CR, Perondi D, Godinho M (2015) Pyrolysis of medium density fibreboard (MDF) wastes in a screw reactor. Energy Convers Manag 92:223–233

Forestry Commission (2010) UK wood production and trade. PDF document, 12 p

Forestry Commission (2014) Forest statistics 2014

Fru C, Irle MA, Maher K, Ormondroyd GA (2003) A protocol for the sampling of recycled wood and resulting chipboard for contamination analysis. In: Proceedings of the seventh European panel products symposium 2003, Llandudno Wales, pp T1-3, 8–10 Oct 2003

Frybort S, Mauritz R, Teischinger A, Muller U (2008) Cement bonded composites—a mechanical review. BioResources 3(2):602–626

Gezer ED, Cooper PA (2009) Factors affecting sodium hypochlorite extraction of CCA from treated wood. Waste Manag 29(12):3009–3013

González-García S, Feijoo G, Widsten P, Kandelbauer A, Zikulnig-Rusch E, Moreira MT (2009) Environmental performance assessment of hardboard manufacture. Int J Life Cycle Assess 14:456–466

González-García S, Lozano RG, Estévez JC, Pascual RC, Moreira T, Gabarrell X, Rieradevall i Pons J, Feijoo G (2012) Environmental assessment and improvement alternatives of a ventilated wooden wall from LCA and DfE perspective. Int J Life Cycle Assess 17:432–443

Gustavsson L, Sathre R (2006) Variability in energy and carbon dioxide balances of wood and concrete building materials. Build Environ 41:940–951

Gustavsson L, Joelsson A, Sathre R (2010) Life cycle primary energy use and carbon emission of an eight-storey wood-framed apartment building. Energy Build 42:230–242

Hagedorn MM, Bockhorn H, Krebs L, Muller U (2003) A comparative kinetic study on the pyrolysis of three different wood species. J Anal Appl Pyrol 68:231–249

Hasan AR, Schneider J, Solo-Gabriele HM, Townsend TG (2011) Online sorting of recovered wood waste by automated XRF-technology. Part I: detection of preservative treated wood waste. Waste Manag 31(4):688–694

Helsen L, Van den Bulck E (2005) Review of disposal technologies for chromated copper arsenate (CCA) treated wood waste, with detailed analyses of thermochemical conversion processes. Environ Pollut 134:301–314

Hill CAS (2011) An introduction to sustainable resource use. EarthScan, London

Hill C, Norton A (2013) Getting the best out of timber—timber products, industrial ecology and carbon sequestration. In: Spear M et al (eds) Proceedings of the international panel products symposium 2013. BioComposites Centre, Wales

Hill C, Norton A, Kutnar A (2015) Environmental impacts of wood composites and legislative obligations. In: Ansell MP (ed) Chapter 12 in wood composites, pp 311–333

HM Government (2009) The UK renewable energy strategy. The Stationery Office, Norwich, 236 p

HM Revenue and Customs (2015) Landfill tax rates. https://www.gov.uk/government/publications/rates-and-allowances-landfill-tax/landfill-tax-rates-from-1-April-2013. Accessed 07 Dec 2015

Huang C, Cooper PA (2000) Cement-bonded particleboards using CCA-treated wood removed from service. Forest Prod J 50(6):49–56

Hurley JW (2004) United Kingdom: national summary reports, COST E31—recovered wood (cited in WRAP 2009 report)

Iida K, Pierman J, Tolaymat T, Townsend T, Wu CY (2004) Control of chromated copper arsenate wood incineration air emissions and ash leaching using sorbent technology. J Environ Eng ASCE 130(2):184–192

Indian Institute of Science (2006) Project completion report on torrefaction of bamboo. http://cgpl.iisc.ernet.in. Accessed 20 Aug 2015

Iritani DR, Silva DAL, Saavedra YMB, Grael PFF, Ometto AR (2015) Sustainable strategies analysis through life cycle assessment: a case study in the furniture industry. J Clean Prod 96:308–318

Irle MA, Ormondroyd GA (2002) A sampling protocol for the determination of contaminant levels in raw wood materials and finished board products. In: Proceedings of the sixth european panel products symposium 2002, Llandudno Wales, pp 154–158, 9–11 Oct 2002

Irle MA, Fru C, Maher K, Ormondroyd GA (2003) Measurement of contaminants in recycled wood and products made from recycled wood. In: Proceedings of the seventh European panel products symposium 2003, Llandudno Wales, pp 208–218, 8–10 Oct 2003

Jacobi G, Solo-Gabriele H, Townsend T, Dubey B (2007) Evaluation methods for sorting CCA-treated wood. Waste Manag 27(11):1617–1625

Jambeck J, Weitz K, Solo-Gabriele H, Townsend T, Thorneloe S (2007) CCA-treated wood disposed in landfills and life-cycle trade-offs with waste-to-energy and MSW landfill disposal. Waste Manag 27(8):S21–S28

Janin A, Zaviska F, Drogui P, Blais JF, Mercier G (2009) Selective recovery of metals in leachate from chromated copper arsenate treated wastes using electrochemical technology and chemical precipitation. Hydrometallurgy 96(4):318–326

Jeffree M (2010) Sustainable construction drive pushes market to timber frame. Timber Trades J. http://www.ttjonline.com/story.asp?storycode=63793. Accessed 14 Mar 2011

John S, Nebel B, Perez N, Buchanan A (2009) In: Environmental impacts of multi-storey buildings using different construction materials. Research Report 2008-02. Christchurch, New Zealand. Department of Civil and Natural Resources Engineering, University of Canterbury

Kamm B, Kamm M (2004) Principles of biorefineries. Appl Microbiol Biotechnol 64(2):137–145

Karade SR (2010) Cement-bonded composites from lignocellulosic wastes. Constr Build Mater 24 (8):1323–1330

Kartal SN, Clausen CA (2001) Leachability and decay resistance of particleboard made from acid extracted and bioremediated CCA-treated wood. Int Biodeterior Biodegradation 47(3):183–191

Kartal SN, Kakitani T, Imamura Y (2004a) Bioremediation of CCA-C treated wood by *Aspergillus niger* fermentation. Holz als Roh- und Werkstoff 62(1):64–68

Kartal SN, Munir E, Kakitani T, Imamura Y (2004b) Bioremediation of CCA-C treated wood by brown-rot fungi Fomitopsis palustris, Coniophora puteana, and Laetiporus sulphureus. J wood Sci 50(2):182–188

Kearley V, Goroyias G (2004) Wood panel recycling at a semi industrial scale. In: Proceedings of the 8th European panel products symposium, Llandudno, pp 1–18, 13–15 Oct 2004

Khan BI, Jambeck J, Solo-Gabriele H, Townsend TG, Cai Y (2006) Release of arsenic to the environment from CCA-treated wood. 2. Leaching and speciation during disposal. Environ Sci Technol 40(3):994–999

Kim G-H, Choi Y-S, Kim J-J (2009) Improving the efficiency of metal removal from CCA-treated wood using brown rot fungi. Environ Technol 30(7):673–679

Klemm D, Kramer F, Moritz S, Lindström T, Ankerfors M, Gray D, Dorris A (2011) Nanocelluloses: a new family of nature-based materials. Angew Chemie Int Ed 50(24):5438–5466

Klemm D, Schumann D, Kramer F, Hassler N, Hornung M, Schmauder H-P, Marsch S (2006) Nanocelluloses as innovative polymers in research and application. In: Advances in polymer science: polysaccharides II, Springer, Berlin, pp 49–96

Kline ED (2004) Southeastern oriented strandboard production. CORRIM phase I final report module E. Life cycle environmental performance of renewable building materials in the context of residential construction. University of Washington, Seattle

Krook J, Martensson A, Eklund M, Libiseller C (2008) Swedish recovered wood waste: linking regulation and contamination. Waste Manag 28(3):638–648

Kumbhar SA, Gupta A, Desai DB (2013) Recycling and reuse of construction waste for sustainable development. OIDA Int J Sustain Dev 6(7):83–92

Lenzen M and Treloar G (2002) Embodied energy in buildings: wood versus concrete—reply to B. Orjesson Gustavsson. Energy Policy 30:249–255

Li J, Brzdekiewicz A, Yang W, Blasiak W (2012) Co-firing based on biomass torrefaction in a pulverized coal boiler with aim of 100 % fuel switching. Appl Energy 99:344–354

Mandel A, Chakrabarty D (2011) Isolation of nanocellulose from waste sugarcane bagasse (SCB) and its characterization. Carbohydr Polym 86(3):1291–1299

Medic D, Darr M, Shah A, Potter B, Zimmermann J (2012) Effects of torrefication process parameters on biomass feedstock upgrading. Fuel 91(1):147–154

Medved S, Irle M, Kržišnik D, Humar M (2015) Partial liquefaction as a method for remediation of recovered wood. In: Proceedings of the international panel products symposium 2015, Llandudno, Wales, pp 97–104, 7–8 Oct 2015

Mitchell A, Stevens G (2009) A life cycle assessment of closed loop MDF recycling using the microrelease process to produce recycled wood fibre from MDF waste. Final report of WRAP project MDD005

Miyatake A, Fiujii T, Hiramatsu Y, Abe H, Tonosaki M (2000) Manufacture of wood strand-cemment composite for structural use. Wood-cement composites in the Asia-Pacific region. Camberra, Australia, pp 148–152

Moghtaderi B, Zanganeh J, Shah K, Wu H (2012) Application of concrete and demolition waste as a CO_2 sorbent in chemical looping gasification of biomass. Energy Fuels 26:2046–2057

Moore N (2010) Improved timber utilisation statistics 2009: consumption of sawn softwood in the main end-user markets in the United Kingdom. Timbertrends report for the Forestry Commission, Oct 2010

Morak A, Unkroth A, Sauerbrey R, Schneider K (1999) Fast analysis of inorganic wood preservatives using laser-induced plasma emission spectrometry. Field Anal Chem Technol 3(3):185–192

Moskal TM, Hahn DW (2002) On-line sorting of wood treated with chromated copper arsenate using laser-induced breakdown spectroscopy. Appl Spectrosc 56(10):1337–1344

Nebel B, Zimmer B, Wegener Z (2006) Life cycle assessment of wood floor coverings. A representative study for the German flooring industry. Int J Life Cycle Assess 11(3):172–182

Nourbakhsh A, Ashori A Ziaei, Tabari H Rezaei (2010) Mechanical and thermo-chemical properties of wood-flour/polypropylene blends. Polym Bull 65(7):691–700

Ntalos G, Papdopoulos A (2006) Mechanical and physical properties of cement bonded OSB. COST action E44–E49 wood resources and panel properties. Valencia, Spain, pp 315–319

Olinzock MA, Landis AE, Saunders CL, Collinge WO, Jones AK, Schaefer LA, Bilec MM (2015) Life cycle assessment use in the North American building community: summary of findings from a 2011/2012 survey. Int J Life Cycle Assess 20:318–331

Optimat (2002) Market study report, UK Timber Frame construction: prospects for home grown timber, 40 p

Pach M, Zanzi Z, Bjornbom E (2002) Torrified biomass a substitute for wood and charcoal. In: Proceedings of 6th Asia-Pacific international symposium on combustion an energy utilization

Papadopoulos AN, Ntalos GA, Kakaras I (2006) Mechanical and physical properties of cement bonded OSB. Holz Roh Werkst 64:517–518

Passer A, Lasvaux S, Allacker K, De Lathauwer D, Spirinckx C, Wittstock B, Kellenberger D, Gschösser F, Wall J, Wallbaum H (2015) Environmental product declarations entering the building sector: critical reflections based on 5 to 10 years' experience in different European countries. Int J Life Cycle Assess 20:1199–1212

Pierre F, Almeida G, Brito JO, Perre P (2011) Influence of torrefaction on some chemical and energy properties of maritime pine and pedunculate oak. BioResources 6(2):1204–1218

Poon CS, Yu ATW, Ng LH (2001) On-site sorting of construction and demolition waste in Hong Kong. Resour Conserv Recycl 32(2):157–172

Puettmann M, Wilson J (2013) Cradle to gate life cycle assessment of composite i-joists production from the southeast. http://www.corrim.org/pubs/reports/2013/phase1_updates/SE%20I-Joists%20July%202013.pdf. Accessed 25 Aug 2015

Qi H, Cooper PA, Wan H (2006) Effect of carbon dioxide injection on production of wood cement composites from waste medium density fibreboard (MDF). Waste Manag 26:509–515

Quaak P, Knoef H, Stassen H (1999) Energy from Biomass: a review of combustion and gasification technologies. World Bank Publications, Washington

Read DJ, Freer-Smith PH, Morrison JIL, Hanley N, West CC, Snowdon P (eds) (2009) Combating climate change—a role for UK forests. An assessment of the potential of the UK's trees and woodlands to mitigate and adapt to climate change. The Stationery Office, Edinburgh, 243 p

Rezaiyan J, Cheremisinoff NP (2005) Gasification technology: a primer for engineers and scientists. CRC Press, Boca Raton

Rincón L, Pérez G, Cabeza LF (2013) Service life of the dwelling stock in Spain. Int J Life Cycle Assess 18:919–925

Rivela B, Hospido A, Moreira MT, Feijoo G (2006) Life cycle inventory of particleboard: a case study in the wood sector. Int J Life Cycle Assess 11:106–113

Rivela B, Moreira MT, Feijoo G (2007) Life cycle inventory of medium density fibreboard. Int J Life Cycle Assess 12(3):143–150

Rosa MF, Medeiros ES, Malmonage JA, Gregorski KS, Wood DF, Mattoso LHC, Glenn G, Orts WJ, Imam SH (2010) Cellulose nanowhiskers from coconut husk fibers: Effect of preparation conditions on their thermal and morphological behavior. Carbohydr Polym 81 (1):83–92

Sandin G, Peters GM, Svanström M (2014) Life cycle assessment of construction materials: the influence of assumptions in end-of-life modelling. Int J Life Cycle Assess 19:723–731

Sarahney H, Wang J, Alshawabkey A (2005) Electrokinetic process for removing Cu, Cr and As from CCA-treated wood. Environ Eng Sci 22(5):642–660

Saravia-Cortez AM, Herva M, Garcia-Dieguez C, Roca E (2013) Assessing environmental sustainability of particleboard production process by ecological footprint. J Clean Prod 52:301–208

Schmidt R, Marsh R, Balatinecz JJ, Cooper PA (1994) Increased wood-cement compatibility of chromate-treated wood. Forest Prod J 44(7/8):44–46

Shi SQ, Gardner DJ, Pendleton D, Hoffard T (2001) Timber production from reclaimed creosote-treated wood piling: economic analysis and quality evaluation. Forest Prod J 51 (11/12):45–50

Siro I, Plackett D (2010) Microfibrillated cellulose and new composite materials: a review. Cellulose 17(3):459–494

Skodras G, Grammelis P, Kakaras E, Sakellaropoulos GP (2004) Evaluation of the environmental impact of waste wood co-utilisation for energy production. Energy 29:2181–2193

Smith RL, Shiau RJ (1998) An industry evaluation of reuse, recycling and reduction of spent CCA wood products. Forest Prod J 48(2):44–48

Solo-Gabriele H, Townsend T (1999) Disposal practices and management alternatives for CCA-treated wood waste. Waste Manage Res 17(5):378–389

Soroushian P, Won J-P, Hassan M (2013) Durability and microstructure analysis of CO_2-cured cement-bonded wood particleboard. Cement Concr Compos 41:34–44

Spear MJ (2015) Preservation, protection and modification of wood composites. In: Ansell MP (ed) Chapter 11 in wood composites, pp 253–310

Spear MJ, Eder A, Carus M (2015) Wood polymer composites. In: Ansell MP (ed) Chapter 10 in wood composites, pp 195–249

Steffen KT, Hatakka A, Hofrichter M (2002) Removal and mineralisation of polycyclic aromatic hydrocarbons by litter-decomposing basideomycetous fungi. Appl Microbiol Biotechnol 60(1–2):212–217

Strom G, Ohgren C, Ankerfors M (2013) Nanocellulose as an additive for foodstuff. Innventia report no. 403. www.innventia.com/Documents/Rapporter/Innventia%20report403.pdf. Accessed 31 Aug 2015

Torvinen K, Sievanen J, Hjelt T, Hellen E (2012) Smooth and flexible filler-nanocellulose composite structure for printed electronics applications. Cellulose 19(3):821–829

TRADA (2002) Part 2: research findings (cited in WRAP 2009 report)

Tumuluru JS, Sokhansanj S, Wright CT, Hess R, Boardman RD (2011) A review on biomass torrefaction process and product properties. In: Proceedings of the symposium on thermochemical conversion. Stillwater, 2 Aug 2001

UK Government (2011) Government review of waste policy in England 2011. https://www.gov.uk/government/uploads/system/uploads/attachment_data/file/69401/pb13540-waste-policy-review110614.pdf. Accessed 10 Dec 2015

UKTFA (2009) UK timber frame association facts and figures. http://www.uktfa.com/#/facts-figures/4538892503. Accessed 14 Mar 2011

US EPA (2006) Solid waste management and greenhouse gases. A life cycle assessment of emissions and sinks, 3rd edn, Sept 2006

Valentin L, Oesch-Kuisma H, Steffen KT, Kähkönen MA, Hatakka A, Tuomela M (2013) Mycoremediation of wood and soil from an old sawmill area contaminated for decades. J Hazard Mater 260:668–675

van der Stelt MJC, Gerhauser H, Kiel JHA., Ptasinski KJ (2011) Biomass upgrading by torrefaction for the production of biofuels: A review. Biomass Bioenergy 35(9):3748–3762

van Faaij A, Ree R, Waldhiem L, Olsson E, Oudhuis A, Wijk A, Daey-Ouwens C, Turkenburg W (1997) Gasification of biomass wastes and residues for electricity production. Biomass Bioenergy 12(6):387–407

Veolia (2015) The circular revolution. Report by Imperial College London for Veolia, lead author, Dr Nick Voulvoulis, 16 p. PDF document. http://www.veolia.co.uk/about-us/about-us/circular-economy/circular-revolution. Accessed 13 Nov 2015

Vick CB, Geimer RL, Wood JE (1996) Flakeboards from recycled CCA-treated southern pine lumber. Forest Prod J 46(11–12):89–91

Vitasari CR, Meindersma GW, de Haan AB (2015) Conceptual process design of an integrated bio-based acetic acid, glycolaldehyde and acetol production in a pyrolysis oil-based biorefinery. Chem Eng Res Des 95:133–143

WBCSD/WRI (2011) Product accounting & reporting standard. www.ghgprotocol.org/files/ghgp/Product%20Life%20Cycle%20Accounting%20and%20Reporting%20Standard.pdf. Accessed 06 Aug 2015

Weatherwax RC, Tarkow H (1964) Effect of wood on the setting of portland cement. Forest Prod J 14(12):567–570

Weber WJ, Jang YC, Townsend TG, Laux S (2002) leachate from land disposed residential construction waste. J Environ Eng ASCE 128(3):237–245

Wei Y, Whou YG, Tomitam B (2000) Hydration behavior of wood-cement based composite I: evaluation of wood species effects on compatibility and strength with ordinary Portland cement. J Wood Sci 46:296–302

Weidema BP, Bauer Ch, Hischier R, Mutel Ch, Nemecek T, Reinhard J, Vadenbo CO, Wernet G (2013) The ecoinvent database: overview and methodology, data quality guideline for the ecoinvent database version 3. www.ecoinvent.org

Werner F, Richter K (2007) Wooden building products in comparative LCA, a literature review. Int J Life Cycle Assess 12(7):470–479

Werner F, Althaus H-J, Richter K, Scholz RW (2007a) Post-consumer waste wood in attributive product LCA; context-specific evaluation of allocation procedures in a functionalistic conception of LCA. Int J Life Cycle Assess 12(3):160–172

Werner F, Althaus H-J, Künniger T, Richter K, Jungbluth N (2007b) Life cycle inventories of wood as fuel and construction material. Final report Ecoinvent 2000 no. 9. EMPA Dübendorf, Swiss Centre for Life Cycle Inventories, Dübendorf, CH

Wicklein B, Kocjan A, Salazar-Alvarez G, Carosio F, Camino G, Antonietti M, Bergstrom L (2015) Thermally insulating and fire retardant anisotropic foams based on nanocellulose and grapheme oxide. Nature 10:277–283

Willeboer W (1998) The Amer demolition wood gasification project. Biomass Bioenergy 15 (3):245–249

Wilson JB, Dancer ER (2004) Laminated veneer lumber—Pacific Northwest and Southeast. In: CORRIM phase I final report module H. http://www.corrim.org/pubs/reports/2005/Phase1/Module_H_Final.pdf. Accessed 25 Aug 2015

Wolfe RW, Gjinolli A (1996) Cement-bonded wood composites as an engineering material In: Proceedings of the use of recycled wood and paper in building applications, Madison, USA

WPIF (2005) WPIF/IG/12.2005 guidance for the specification and control of post-consumer reclaimed wood raw materials used in the manufacture in the UK of wood particleboard, MDF and OSB

WRA (2008) Wood Recyclers' Association waste wood market statistics 2007. http://www.woodrecyclers.org. Accessed 21 Feb 2011

WRA (2010) Wood Recyclers' Association, waste wood to markets statistics 2009 (in thousand tonnes). http://www.woodrecyclers.org/documents/stats2009.pdf. Accessed 21 Feb 2011

WRA (2015) Wood Recyclers' Association. http://www.woodrecyclers.org/about-wood-recycling/intro/. Accessed 18 Nov 2015

WRAP (2008) Construction, demolition, and excavation waste arisings, use and disposal for England 2008. Final Report CON900-001, 108 p

WRAP (2009) Wood waste market in the UK, Summary report, Project code MKN022, Poyry Forest Industry Consulting Ltd and Oxford Economics Ltd, 37 p. Accessed 29 Aug 2015

WRAP (2013) Energy from waste (EfW) plant database. Spreadsheet, 1 p. http://www.wrap.org.uk/content/list-energy-waste-sites. Accessed 13 Nov 2015

WRAP (2015) Guidance on separating wood for recycling at source. PDF document, 5 p. http://www.wrap.org.uk/content/guidance-separating-wood-recycling-source-0. Accessed 13 Nov 2015

WRAP & MEL Research (2005) Review of wood waste arisings and management in the UK (cited in WRAP 2009 report)

Ximenes FA, Grant T (2013) Quantifying the greenhouse benefits of the use of wood products in two popular house designs in Sydney, Australia. Int J Life Cycle Assess 18:891–908

Xiong R, Zhang X, Tian D, Zhou Z, Lu C (2012) Comparing microcrystalline with spherical nanocrystalline cellulose from waste cotton fabrics. Cellulose 19(4):1189–1198

Xu X, Jayaramana K, Morinb C, Pecqueuxb N (2008) Life cycle assessment of wood-fibre-reinforced polypropylene composites. J Mater Process Technol 198:168–177

Yasuda K, Tanaka M, Deguchi Y (2006) Basic evaluation of sorting technologies for CCA treated wood waste. In: WIT transactions on biomedicine and health, Vol 10. Wessex Institute of Technology, UK pp 319–327

Yorulmaz SY, Atimtay AT (2009) Investigation of combustion kinetics of treated and untreated waste wood samples with thermogravimetric analysis. Fuel Process Technol 90:939–946

Zhou Y, Kamdem DP (2002) Effect of cement/wood ration on the properties of cement-bonded particleboard using CCA-treated wood removed from service. Forest Prod J 52(3):77–81

Additives in Wood Products—Today and Future Development

Dick Sandberg

Abstract Most wood products include additives. They may be preservatives to protect the wood against biological degradation or against fire, coatings for protection or to give the wood a more favourable aesthetic appearance, non-wood materials to improve the performance of the product and overcome weaknesses in the wood material, or plastics in combinations with wood residues to create new types of wood–plastic combinations. The global wood industry is, for example the largest user of adhesives; about 80 % of all wood and wood-based products involve some form of bonding and 70 % of the total volume of adhesives produced is consumed in the woodworking industry. Wood can thus be regarded as a composite consisting of wood-based materials combined with other materials to form an aggregate material. An example is plywood, in which veneers are joined with adhesive to form a flat panel. Other types of wood composites include various board products, structural composite timber and, furniture and joinery components, all including some form of bonding with adhesive. This situation obviously influences the way in which we should relate to wood products and their environmental impacts. This chapter gives a state-of-the-art presentation of different additives currently being used in wood products. This information is necessary for further studies on the influence that these additives have on the service life and on environmental aspects, and the limitations which they may impose on the reuse, recycling and upgrading of wood products.

Keywords Adhesives · Chips · Engineered wood products · Fibres · Modified wood · Particles · Strands · Surface treatments · Veneer

D. Sandberg (✉)
Wood Science and Engineering, Luleå University of Technology,
931 87 Skellefteå, Sweden
e-mail: dick.sandberg@ltu.se

A. Kutnar and S.S. Muthu (eds.), *Environmental Impacts of Traditional and Innovative Forest-based Bioproducts*, Environmental Footprints and Eco-design of Products and Processes, DOI 10.1007/978-981-10-0655-5_4

1 Introduction

Since the dawn of civilization, wood has been used in its natural state because of its unique advantages: widespread availability, natural renewal, favourable ecological assessment and flexibility of implementation. Although in the last 150 years we have witnessed an ever-increasing dependence on steel and concrete in structural applications, wood remains an important feature of our infrastructure and one of the few truly renewable resources available to us. Only recently has wood been developed to form a range of products that are increasingly functional, based on a combination of performance and sustainability requirements. This has been possible because of new industrial processes which extend the size and modify the properties of natural wood, and the need to use manufacturing residues and lower grade trees to produce more versatile and more consistent products. The result is a vast array of materials known as engineered wood products (EWPs).

To transform wood into an industrially manufactured engineering material it is necessary in a refining process to combine this natural material with other materials or substances with a broad range of origins, here referred to as additives. This chapter gives an introduction to the additives commonly used in the wood mechanical industry, i.e. the industry which turns the forest into sawn timber, semi-finished products, packaging, construction wood, furniture and interior fittings and in the board industry. The focus is on substances added to the wood either to bond pieces of wood together to change their dimensions or shape or to improve the properties of the wood. Some practical aspects on recycling and the combustion of wood products close the chapter. First, however, an introduction is given to EWPs and the additives that are used in these products.

2 Additives in EWPs

EWPs, also referred to as reconstituted wood, wood-based products or wood-based composites, are wood components in general and structural components for industrial use in the production of furniture, in interior and exterior joinery and in building construction. They have in common well-defined dimensions and properties that satisfy a particular need. This family of wood products differs from ordinary sawn timber in the way in which it is further processed to create the "engineering properties". This apparently means that stress-graded sawn timber should be included in this group of wood products, but this is not the case. Wood composites have in general poorer mechanical properties than sawn timber in the longitudinal direction but, because of their more consistent properties in both planar directions and in the cross section of the composite, the safety margins can be kept narrower than when sawn timber is used.

For simplicity in this chapter, this impressive range of wood products is simply designated EWPs. EWPs are a practical way to achieve large structural member sizes, far beyond those available with single pieces of timber. In addition, such composite materials can be engineering to be much stronger, stiffer and more dimensionally stable than solid wood by dividing and randomizing defects, densifying the material and strategically placing higher and lower quality materials in high and low stress locations, respectively. With imaginative design and the use of engineered materials, it is possible to create large wooden buildings, for industrial, institutional and residential uses.

Wood itself is a cellular biopolymer composite consisting of cellulose, hemicelluloses, lignin, extractives and inorganics. The three macromolecular cell-wall components of wood play specific roles in determining the properties of the wood cell wall. The cellulose microfibril provides the tensile reinforcement of the wood cell wall, with an exceedingly high tensile modulus of elasticity, of the order of 145 GPa. However, although strong in tension, cellulose microfibrils buckle easily when subjected to a compressive load. Lignin provides a rigid enveloping matrix for the microfibril and provides resistance to compressive loads. The surface of the microfibril is highly polar, with a high density of hydroxyl groups on the surface, but the lignin matrix has a much lower density of hydroxyl groups, and this leads to a low adhesive interaction and hence poor interfacial stress transfer between the microfibril and lignin. One role of the hemicelluloses is to act as an interfacial coupling agent between the surface of the microfibril and the lignin matrix.

Nature is programmed to recycle wood back into its basic components of carbon dioxide and water through various degradation mechanisms, including biological, photochemical, thermal, aqueous, chemical and mechanical degradation. To customize or improve its natural properties the better to fit the environment in which it will be used, wood is often combined with natural or synthetic substances such as adhesives, coatings and preservatives or is mechanically joined with nails, screws, bolts, snaps, notched metal strips, etc. For example, glued laminated timber (GLT) has been used for approximately a century to give constructions with properties better than those of sawn timber. Other structural wood-based materials are wood composites, where wood is used in a fractionated form. The application of a coating is the best-known way of changing the characteristics of a wood surface, with regard to both appearance and its mechanical properties such as hardness and abrasive resistance.

One of the key features in wood processing is to change the dimensions of the wood material from "tree dimensions" to an engineered material with properties and dimensions adapted to the purpose for which it is to be used. Figure 1 shows the principal ways that are currently being used in industry to convert trees to wood products. After the forest is harvested, the trees are sorted into different classes depending on their intended use. To produce wood for construction purposes, sawmilling is today the dominant process, yielding sawn timber in well-defined dimensions as well as by-products such as bark, sawdust and chips. An alternative process is the production of veneers for subsequent use for board manufacture (plywood), elements for construction purposes (laminated veneer lumber) or the manufacture of moulded products (laminated veneer products) and high-density materials for interior and special applications (high-pressure laminated veneer).

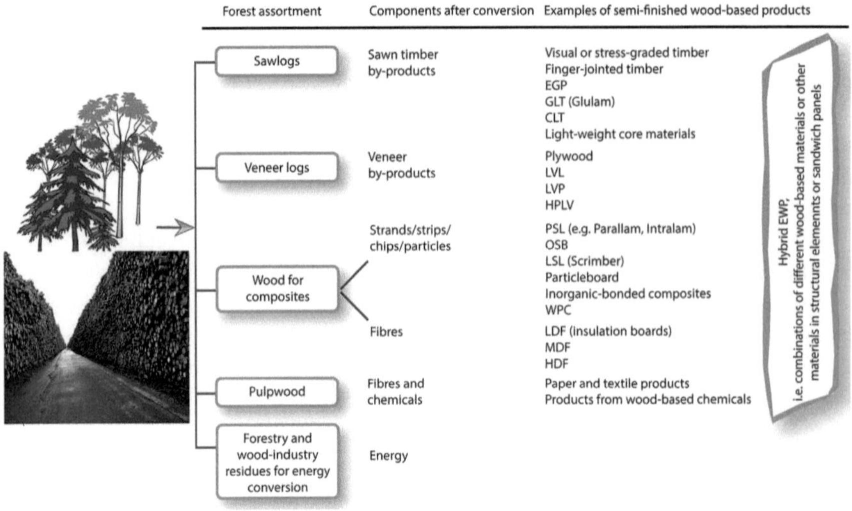

Fig. 1 The industrial use of the forest resource. Depending on species, dimensions and categories, the forest raw material is after harvesting sorted into different classes according to their industrial use. *EGP* edge-glued panels, *GLT* glued-laminated timber, *CLT* cross-laminated timber, *LVL* laminated veneer lumber, *LVP* laminated veneer products (moulded products), *HPLV* high-pressure laminated veneers, *LDF*/*MDF*/*HDF* low/medium/high-density fibreboards, *PSL* parallel strand lumber, *OSB* oriented strand boards, *LSL* laminated scrimbed lumber, *WPC* wood plastic composites

After the transformation of the tree in the sawmill, the sawn timber does not usually have the dimensions required in the final product, and a lot of effort is being made in the wood industry to transform the sawn timber into dimensions and grades that suit the products requested by the consumers. Joining wood is a major step in these processes, and adhesives of different types are here as the key component.

Trees or parts of trees that are not suitable for use in the sawmill or in veneer processes have, if used for industrial purposes at all, three main uses: for paper pulp production, for the manufacture of wood-based composites or for energy conversion. These processes also use the by-products from the sawmill and veneer processes or residues from other woodworking industries, and in some cases also agricultural waste. The board industry produces a variety of wood-based panel products (oriented strand board, flake board, particleboard, hardboard, insulation board, medium-density fibreboard, cement-bonded board, etc.) based on comminuted wood in different sizes from long and thin flakes (veneer flakes) to fibre bundles that are commonly bonded together by an adhesive or by integral bonding achieved by interfelting of the fibres and in some cases by a ligneous bond. Other materials may be added to improve certain board properties. Wood plastic composites (WPC) are a rather new building material on the market, based on a thermoplastic matrix and a wood component. The matrix is usually recycled polyethylene or polypropylene, and the wood is sawdust or shaving residues from the wood industry (Carus and Gahle 2008).

EWPs also include structural elements where semi-finished wood-based components are combined to form products such as I-joists, in some cases in combination with other materials. The large variety of sandwich panels is other examples of these hybrid EWPs that can be found on the market, where the engineered properties or functions are even more specific than those of the semi-finished EWPs. Due to the almost unmanageable number of varieties of hybrid EWPs, we shall not provide any deep description of these materials in this section.

The pulp and paper processes and energy conversion based on forestry and wood industry residues are outside the scope of this presentation and are omitted. For more basic studies of these subjects, further reading in Ek et al. (2000) and Hood et al. (2011) is recommended.

A characteristic of EWPs is that, regardless of the dimensions and shape of the wood raw material (from fibres to sawn timber), the wood in a EWP almost always bonded together with an adhesive. Only in the case of insulating boards and fibreboards manufactured by the wet process and self-bonded veneer boards is no adhesive or other binder added in the production process. The variety of properties of EWPs is very large and it is beyond the scope of this survey to give a detailed description them all. Below is a brief description of the most common EWPs (according to Fig. 1) and fibre-reinforced wood with a focus on the additives.

2.1 EWPs Based on Sawn Timber

EWPs based on sawn timber are lengthwise joined sawn timber, edge-glued panels (EGP), glued-laminated timber (GLT) and cross-laminated timber (CLT).

2.1.1 Lengthwise Joining of Sawn Timber

Lengthwise joining of sawn timber is performed for several reasons, the most common being: (1) to make use of wood waste from, e.g. length adaptation of sawn timber, (2) to achieve length longer than that of the log and (3) to remove defects and increase the strength or to influence the appearance of the sawn timber. Because of the small surface area of the cut surface (the cross-sectional area of the timber) only low tensile forces can be transmitted by simply butt-jointing the ends of the sawn timber. Different methods for joining sawn timber have therefore been developed, and Fig. 2 shows some of them. Today, the most common method for joining sawn timber industrially is finger-jointing.

Finger-jointed timber is used for structural members such as studs and in glulam (GLT), as well as for non-structural purposes. In species in which short clear lengths of wood are separated by knot whorls, it occurs that all defects, mostly knots, are cut away and the clear pieces of sawn timber are finger-jointed to specific ready-to-use components.

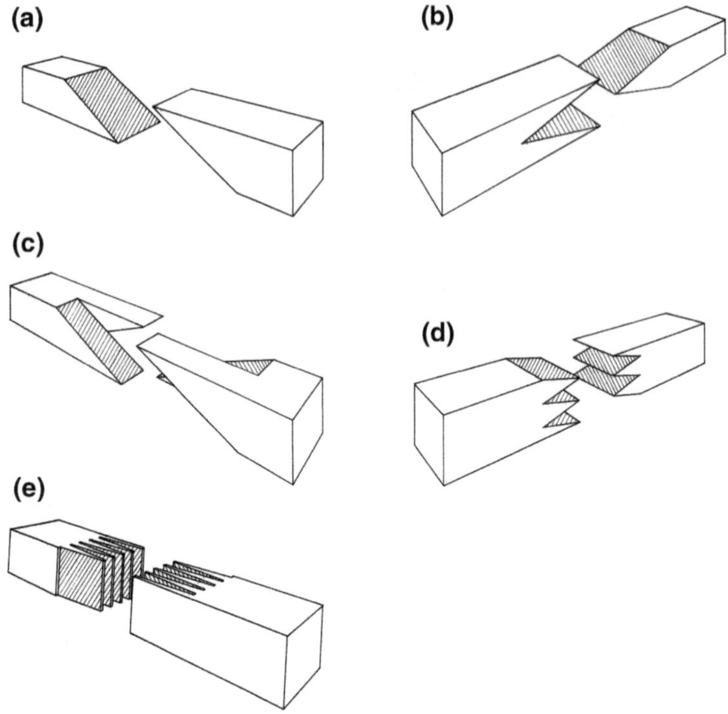

Fig. 2 Different principles for lengthwise joining of sawn timber: (**a**) simple scarf joint, (**b**) wedge tenon, (**c**) "American" scarf joint, (**d**) multiple wedge tenon and (**e**) finger-joint

The choice of adhesive in finger-jointing depends on the final use of the product. For structural purposes phenol-resorcinol-formaldehyde (PRF) or melamine-urea-formaldehyde (MUF) types are commonly used, and for non-structural products polyvinylacetate (PVAc) or hot melting adhesive are used, but other types of adhesives are also frequently used for joining sawn timber in the industry.

2.1.2 EGP—Edge-Glued Panels

An edge-glued panel is a board of widthwise glued lamellae of solid wood. The lamellae are sometimes joined lengthwise, especially in panels of lower grade timber. EGPs are used for furniture, for exterior and interior joinery purposes, or as core for sandwich panels with outer sheets of other wood or non-wood material. The lamellae and the outer sheet are glued with adhesives of various types depending on the properties asked for in the application. EGPs are produced with three levels of adhesive bond quality: (1) interior adhesives that are non-moisture resistant, (2) intermediate moisture resistant adhesives, i.e. lower resistance to moisture than exterior but greater than interior adhesives and (3) exterior adhesives that are moisture resistant. Minimum requirements for each type, developed from the results of long-term exposures, are detailed in product standards for EGPs. The most common adhesives in EGPs for interior usage are different types of PVAc adhesives. The amount of adhesive per unit area of bond-line surface is 100–200 g/m^2. The manufacture of EGPs ranges from manual carpenter-made panels to highly industrialized processes in specialized production units.

2.1.3 GLT—Glued-Laminated Timber

GLT, commonly referred to as glulam, is a structural element made of sawn timber glued together with parallel fibre orientation to straight beams or members with some kind of curvatures. In Europe, and also in other parts of the world, GLT is used in a wide variety of applications, ranging from supporting beams in residential framing to major structural elements in non-residential buildings as girders, columns and truss members. Softwood and hardwood species are being used in GLT, but the most common species for GLT production are Norway spruce, Douglas fir, larch, Scots pine, southern pine, radiata pine and yellow poplar. For structures that are expected to be exposed to the prolonged influence of moisture, preservative-treated sawn timber is used. The major advantage of GLT is its high strength and stiffness, which makes it possible to manufacture structures for wide spans and enhanced bearing capacity.

The manufacture of GLT is very similar, regardless of factory and country. Figure 3 shows the principal stages in the manufacture of GLT. Stress-graded sawn timber is general. The cross section of the GLT can be built up from lamellae with approximately the same strength, so-called homogeneous glulam. In order to utilize the timber's strength in the best way, high-strength sawn timber is used for the outer parts of the beam where the stresses are highest, so-called combined glulam.

In GLT manufacture, only adhesives that have a documented high strength and durability under long-term loads are used, and only those adhesive for which the producers have a long practical experience. The different adhesives that can be used for GLT are well regulated and under constant control. In Europe, the formal requirements are given by the European standard EN-301 (CEN 2013) which classifies two types of adhesives, type I and II. The type I adhesives can be used in all climates, while adhesive of type II have restrictions limiting where they can be used.

The traditional and very common adhesives used in the manufacture of GLT are those of the synthetic two-component PRF type (phenol-resorcinol-formaldehyde). All PRF adhesives used for GLT production are of type I. PRF adhesives give a dark reddish-brown bond line.

MUF (melamine-urea-formaldehyde) adhesives are now being increasingly used. MUF adhesives also being type I. The bond lines of MUF adhesives are initially bright but darken over time.

For finger-jointing of lamellae, PFR or MUF adhesives are used. Finger joints can appear as dark spots or thin lines on the GLT surfaces. The labelling of the GLT will specify the type of adhesive used in the production, i.e. type I or II according to EN-301 (CEN 2013).

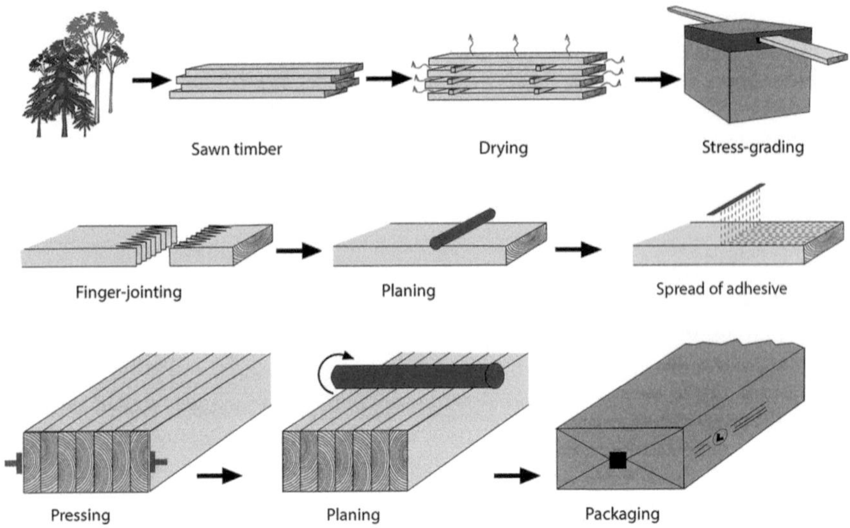

Fig. 3 The main steps in manufacture of glued-laminated timber (GLT)

There is a continuous development of adhesives for GLT, and new adhesives are gradually being introduced. An example of an adhesive recently approved is the single-component polyurethane (PU) adhesive (type II).

2.1.4 CLT—Cross-Laminated Timber

CLT, also called X-lam, is a straight or curved, multi-layer timber member consisting of at least three layers, of which at least one of three are orthogonally bonded, which always include timber layers and may also include wood-based panel layers. CLT is a material for the manufacture of structural elements for use in buildings and bridges. CLT can be multi-layered boards normally with a thickness around 70 mm and consisting of 3, 5 or 7 layers or as CLT multi-layered blocks with a thickness of up to 600 mm.

2.1.5 Lightweight Core Materials

Lightweight materials based on wood for interior fittings and furniture have been of interest for at least the last 50 years, mainly for cost-reducing reasons. Today, the increasing care of the environment and the growing interest in the concept of a sustainable society provide further impulses for the development of lightweight materials. The main benefits of the lightweight materials for furniture applications are their high strength-to-density ratio. A low weight is advantageous in the transport and handling of products and it also lowers the transportation costs.

In terms of recycling and resource-efficient material management, lightweight materials made of wood, paper or recycled cellulosic materials have been developed in recent years mainly in the furniture industry. Feifel et al. (2013) have shown that an increased used of lightweight wood-based materials for furniture purposes in general can greatly lower the environmental impact of greenhouse gas emissions, non-methane volatile organic compounds (NMVOC) and formaldehyde from that industry. Lightweight structural materials based on wood and paper layers are being increasingly considered as substitutes for commonly used materials such as fibreboards or particleboards. In terms of recycling and energy consumption, foam and composite cores based on non-renewable resources are, however, critical.

Lightweight materials can be divided according to their function and structure into three groups: (1) *lightweight materials* which combine materials with a low weight-to-strength/stiffness ratio, (2) *structural lightweight materials* which are structures mainly for building purposes that, with a minimum of weight, can distribute applied loads and (3) *lightweight systems*—the superposition of functions of lightweight materials, providing not only supporting functions but also thermal insulation, etc.

During recent years, several ideas for lightweight panels of exclusively solid wood that combine lower weight with high strength have been presented. These panels are mainly for buildings and other structural systems due to their wide range of advantages, i.e. high strength-to-weight ratio, reduction in the cost of framework and good thermal and sound insulation properties (Pokharel 2003). DendroLight is a core material that is available in the market that consists of layers of cross-aligned sawn timber with longitudinal kerfs, which are glued to the surface layer at an angle of 45° (Berger 2006). Skuratov (2010) has presented a three-layer panel with a core of low-grade sawn timber that has a pattern of hollow cells which reduce its weight. A lightweight panel for the use in furniture construction, for door blanks and for construction of internal wall elements was suggested by the group around Prof. Dick Sandberg in the beginning of the twenty-first century, and is shown in Fig. 4 (Näsström 2005).

2.1.6 Fibre-Reinforced Wood

Wood members reinforced with a fibre-reinforced polymer (FRP) composite systems are used in a variety of structural applications, such as GLT for construction (Hallström 1995), GLT for bridges (Dagher et al. 2002; Lopez-Anido and Xu 2002), reinforced railroad ties (Sonti and GangaRao 1996) and the repair of wood piles (Lopez-Anido et al. 2003). Plywood can also be reinforced with fabric, fibres, etc., between the veneer layers. According to Lopez-Anido et al. (2005) there are three procedures to reinforce wood members with FRP composites: (a) wet lay-up of the fabric reinforcement on wood members, (b) adhesive bonding of prefabricated sheets to wood members and c) adhesive infusion, i.e. a variation of the vacuum-assisted adhesive transfer moulding (VARTM) process that applies the Seemann composite adhesive infusion moulding process (SCRIMP), Table 1.

Fig. 4 Example of a lightweight panel manufactured of solid wood. The core elements are cross-laminated to the surface panels. The holes in the core elements are for efficient assembly

Table 1 Hybrid FRP-GLT composite fabrication systems (Lopez-Anido et al. 2005)

System	Advantages	Concerns
Wet lay-up of fabric	Use of impregnator Flexible Ease of use in restricted areas Compaction with vacuum	Quality control Entrapped air Compaction and fibre wrinkling Environmental issues
Adhesive bonding of prefabricated sheets	Reinforcement prefabrication Adhesive bonding Rapid procedure Ease of fabrication	Shear lag effect Durability of the adhesive Two-step fabrication
Resin infusion (VARTM/SCRIMP)	Placement of dry fabric Infusion under vacuum High compaction Fill wood cracks	Fibre wet-out Difficulty of holding vacuum Vacuum bag, flow media, and conduits

Table 2 General mechanical properties of E-glass fibres (Including E-glass with and without boron oxide)

Property	
Density (kg/m^3)	2500–2600
Tensile strength, filament (GPa)	3.1–3.8
Tensile modulus, filament (GPa)	75–80
Ultimate strain, filament (%)	4.5–4.9
Poisson's ratio	0.18

The most common reinforcement fibres for wood are glass fibres, but for special applications carbon and aramid fibres can be used. Shade (1998) refers to carbon and aramid fibre reinforcement as "overkill" for wood applications having a very high specific strength and stiffness.

The most suitable reinforcement for a wood construction material is biaxial E-glass because of the orientation of the fibres in two plies (Shade 1998). The layers may be displaced parallel or at an angle of 45° to the length of the fabric. E-glass fibres are available also in non-woven fabric. E-glass refers to fibreglass produced from a common alumina–borosilicate glass. E-glass composition offers high strength, stiffness, corrosion resistance, low electrical conductivity and essentially isotropic properties. The properties of the E-glass may vary mostly due to a variation in the boron oxide content. The latest commercial needs make it necessary to reduce the boron oxide content in the E-glass composition. Some of the mechanical properties of the E-glass fibres are presented in Table 2.

Thermosetting composite materials form the matrix of the reinforcement. Composites based on thermosetting resins are created when the monomer or pre-polymer liquid is transformed through a chemical reaction into a cross-linked polymer (Wallenberger and Bingham 2010). Thermosetting adhesives are infusible, hard and brittle compared to thermoplastic composites. The requirements of the thermosetting matrix resin used for wood reinforcement are to ensure strong adhesion with a wooden core; to ensure good compensative compression strength for the FRP composite; low weight, reduced costs, a low water absorption rate and reduced environmental hazards. Typical adhesives are unsaturated polyesters, epoxies, vinyl esters, phenolics, PU and silicones.

The fillers are additives in the matrix with a cost reduction purpose. However, some fillers may yield special composite properties. Inorganic fillers such as calcium carbonates, hollow glass spheres and wollastonite are widely used with unsaturated polyester resins.

Release agents are used to prevent a moulded composite from adhering to the surface when curing, i.e. gluing to the pressing moulds. Most popular release agents are sprayed onto the mould surface as a liquid (e.g. silicones), wax or solid film. Polyethylene film can serve as release agent. Release agents can sometimes be mixed with the adhesive to enhance the processing. Common property of the release agents is a low surface energy that hinders the adhesion of the matrix to the mould. The final products will, in this case, contain these substances.

Fig. 5 Moulded tube from Norway spruce with a length of 250 cm, a diameter of 28 cm, and a thickness of the tube wall of 2 cm, fibre reinforcements: (*middle*) filament winding of aramid and carbon, (*right*)

Haller et al. (2013a) suggested a thermo–hydro–mechanical process to manufacture wooden tubes out of sawn timber for load-bearing and conveying applications and they described the implications for construction and the environment, Fig. 4. The tubes were fibre reinforced at the outer face to increase strength and protect the structure. Haller (2007) emphasized that a high load-bearing capacity and a large saving of time and material can be achieved if EWPs are used instead of sawn timber. Haller et al. (2013b) also investigated the use of moulded tubes in aggressive environments. The tubes were fibre reinforced at the outer face, Fig. 5.

As a medium, hot, highly concentrated brine was conveyed at temperatures of up to 60 °C. Compared to steel tubes, the spruce tubes did not show any noticeable erosion after 4 weeks exposure, whereas the steel tubes had been considerably affected. To connect the tubes, fittings made of resin-impregnated Compreg (see Sect. 2.2.4), specifically designed for the purpose were used. In Kutnar et al. (2015) it is shown how the moulded and fibre-reinforced tubes are used for wind turbines, Fig. 6.

Fig. 6 Wind-power plant from fibre-reinforced moulded tubes

2.2 EWPs Based on Veneers

EWPs based on veneers are plywood, laminated veneer lumber (LVL), laminated veneer products (LVP) and high-pressure laminated veneer. They are briefly presented below.

2.2.1 Plywood

Plywood is a rigid board composed of an odd number of veneers glued together so that the fibre orientation of the veneer is perpendicular to the fibre orientation of the adjacent veneers. Plywood mills cut veneer for their own production and, after turning, the sheets are cut into the required dimensions. The sheets are dried and glued together in a hot press under high pressure, and are then trimmed, polished and conditioned to the desired moisture content. There are numerous grades, but plywood can be divided into plywood for constructional (exterior) purposes, for interior use (joinery and decorative plywood) and for special application such as concrete shuttering, marine plywood, plywood with special surface layers, etc. Board types where veneer sheets are glued to a timber-strip core, i.e. a core plywood or face-glued blockboard, are also assigned to the group of plywood.

Additives in plywood are:

- Adhesives for bonding the veneers
- Surface layers or treatments

Plywood is glued with a thermosetting adhesive. Phenol-formaldehyde-based adhesives are used for exterior-type plywood and urea-formaldehyde (UF), reinforced urea adhesives and sometimes natural polyphenols (tannins) mixed with synthetic adhesives are used for interior-type plywood. The spread (amount of adhesive per unit area of bond-line surface) varies from about 100 to 500 g/m^2 depending on several factors pertaining to the wood, adhesive and manner of application.

Plywood panels may be surfaced with metals, plastics, or other material, or their veneers may be impregnated to achieve a superficial hardness, or resistance to micro-organisms, fire or other destructive agents.

2.2.2 LVL—Laminated Veneer Lumber

Laminated veneer lumber (LVL and also Microllam LVL) is made of veneers with a thickness between 2 and 4 mm bonded together with the same fibre orientation in the layers. The manufacturing procedure is very similar to that of plywood production. The different stages in the manufacture of LVL are intended to eliminate the defects inherent in ordinary sawn timber and to develop a higher strength, a good dimensional stability and a homogeneity of physical and mechanical properties along the beam or elements. LVL is in most cases bonded with a phenol-formaldehyde (PF) or melamine-formaldehyde (MF) adhesive, and has a higher strength and stiffness than plywood. LVL can be found in a large variety of dimensions; length up to 25 m, thicknesses below 150 mm and width up to 2.5 m.

2.2.3 LVP—Laminated Veneer Products

LVP consist of veneers bonded together with an adhesive under pressure into a predetermined shape and usually under an elevated temperature to decrease the curing time of the adhesive. The process used to manufacture such products is generally called laminated bending and it is commonly used for the manufacture of components for exterior and interior use.

Different techniques for forming LVPs have been developed during the last hundred years: male and female moulds, pressing lamination to shape by means of metal tension bands or with an inflated flexible rubber hose and a metal strap, etc. Stevens and Turner (1970) give a good overview of the various techniques for forming LVP. The most frequently used industrial method for bending laminated assemblies to the desired shape and applying the requisite pressure to achieve proper bonding together of the laminae is to press the assembly in a mould between shaped male and female forms (Navi and Sandberg 2012).

The adhesive is extremely important for the function of the final LVP, and the development of synthetic adhesives has contributed to their development. Common adhesives for the lamination of veneers are based on reactions of formaldehyde with phenol, resorcinol, urea, melamine or a mixture thereof (Rowell 2005). Urea-formaldehyde (UF) adhesives; melamine urea-formaldehyde (MUF) adhesives or emulsion; polymer isocyanate (EPI) adhesives; PU adhesives; and epoxy adhesives are examples of adhesives used for the laminated bending of veneers. Urea-formaldehyde-based adhesives are the adhesives most frequently used in the industry for LVP.

2.2.4 HPLV—High-Pressure Laminated Veneer, DW—Densified Wood

High-pressure laminated veneer (HPLV) and densified wood (DW) are wood materials that have been compressed in the transverse direction in order to increase their density. Compression in the transverse direction reduces the void volume of the lumens in the wood material and increases its density. This process is commonly called densification (Sandberg et al. 2013). One of the reasons for densifying wood in the transverse direction is to produce high-quality components from timbers of low quality. However, densified wood has an undesirable property, i.e. a tendency to recover all or part of its compression set and return to its initial dimensions when subjected to heat and humidity. In most of densification processes that have been industrialized, this recovery problem has been solved by the development of adhesive-impregnated products, which are now being commercially produced. Some examples of products that can be found on the market are presented here.

Solid wood treated with a thermosetting, fibre-penetrating adhesive and cured without compression is known as Impreg. The wood is soaked in the aqueous adhesive solution or, if air dry, is impregnated with the solution under pressure until the adhesive content amounts to 25–35 % of the weight of dry wood. The treated wood is allowed to stand under non-drying conditions for 1–2 days to permit a uniform distribution of the solution throughout the wood, and the adhesive-containing wood is then dried at a moderate temperature to remove the water and finally heated at a temperature of 60–150 °C to cure the adhesive. Very high drying rates should be avoided to prevent excessive migration of the adhesive to the surface. A number of different adhesive-forming systems have been successfully polymerized within the cell walls of wood, namely phenol, resorcinol, melamine- and urea-formaldehydes, phenol furfural, furfuryl aniline and furfuryl alcohol (Stamm and Seborg 1939).

Compreg is similar to Impreg except that it is compressed before the adhesive is cured within the wood. The adhesive-forming chemicals (usually PF) act as plasticizers for the wood so that it can be compressed to a density of 1200–1350 kg/m^3 at a pressure of about 7 MPa and a temperature of 125–150 °C (Kollmann et al. 1975).

Electrical transmission support components made from densified wood are typically adhesive-impregnated laminated veneer (Kamke 2013). Low molecular weight adhesives (typically PF) are used to impregnate the veneer, which is then partially cured in an oven. The impregnated wood is compressed in a heated press (open system) to a density of approximately 1300 kg/m^3. Another use for adhesive-impregnated densified veneer is in storage containers for liquid natural gas and associated support structures (wear plates for machinery and transportation vehicles, machine pattern moulds, bullet-proof barriers and some structural building components).

Panzerholz compressed wood (Delignit), is an extremely hard multi-layered sheet material produced in Germany. It is manufactured from beech veneers and synthetic adhesives under heat and a high pressure. Under these conditions the veneers are compressed to half their original thickness. The material is used for security panels, tooling, jigs, moulds, transformer parts, support for liquid natural gas tanks, neutron shielding and audio component cases.

Another product produced in Germany by Deutsche Holzveredelung is Dehonit, a compressed laminated wood manufactured from high-quality selected beech veneers coated or impregnated with a special synthetic phenolic adhesive. It is pressed under high pressure and temperature to form a laminate material.

In France, Permawood, also known as Lignostone, is a laminated densified wood according to the DIN 7707 and IEC 61061 standards (DIN 1979; IEC 2006), made of beech veneers laminated together using synthetic adhesive that harden under pressure and heat.

In the United States, densified wood is on the market under the trade names Permali and Insulam. These materials are densified, phenolic impregnated and laminated products made from beech veneers, and are laminated with cross-directional fibres. The veneers are impregnated under vacuum and then densified through the application of heat and pressure. The result is a homogeneous material with high strength and toughness and excellent dimensional stability and dielectric properties. The product is used for electric power equipment, structural supports in cryoenvironments and electrical insulation for rail transportation vehicles.

In Australia, Insulcul Services Ltd. is producing densified wood manufactured from beech veneers, impregnated with a synthetic adhesive and densified under high pressure and temperature. The veneers are compressed at 90 °C, resulting in a material with uniform strength and stability. The product is used by the electrical power industry.

In Italy, a laminated and densified beech products (RANPREX) is produced by Rancan Srl. The product is impregnated with special thermosetting adhesive and densified at high pressure and temperature. The product is an excellent electrical and thermal insulator, with good physical and mechanical characteristics under compression and bending loads and with good impact resistance. The material is also self-lubricating and extremely resistant to wear. It is used for many different applications; it is used in the electrical power distribution industry, for support beams, treaded rods, compression blocks and pressure rings.

Adhesive-impregnated laminated densified wood products are also produced in India by, e.g. Surendra Composites Private Ltd. They produce laminates of adhesive-impregnated veneer, which are used for electrical power transmission equipment and machine parts.

Olympus Corp. in Japan has developed a three-dimensional moulding process for wooden materials in which the density of a piece of cypress wood can be increased from approximately 450 kg/m^3 to more than 1000 kg/m^3. The resulting material is thin enough to be used as a casing material for electronic products, but is much harder than, e.g. ABS plastics and polycarbonate-adhesive-based engineering plastics that are normally used in such applications.

MyWood2 Corporation (Iwakura, Aichi, Japan) manufactures densified solid cedar wood products. Their primary market is flooring in Japan and China and the products are also sold for use in furniture. The MyWood2 product is wood impregnated with a polymer to provide resistance to water, and compressed to approximately 50 % of its original volume.

2.3 *EWPs Based on Strands, Strips, Chips and Particles*

EWPs based on strands, strips, chips and particles include parallel strand lumber, oriented strand boards, laminated scrimbed lumber, particleboards, inorganic-bonded composites and wood–plastic composites.

2.3.1 PSL—Parallel Strand Lumber

The manufacture of parallel strand lumber (PSL, also called Parallam) is based on a technology which makes it possible to convert small trees into elements with large cross sections (up to ca. 30 × 50 cm^2) and considerable lengths (up to 20 m). These products are intended for building construction for elements in compression, large trusses, beams or posts.

PSL is manufactured from parallel-oriented veneer strips, made from rotary cut veneer or veneer waste. After drying to a moisture content of about 6 %, the strands are treated with an adhesive with hydrophobic properties and introduced in the longitudinal direction into a continuous press where the adhesive is polymerized by microwaves. After sanding, the PSL is cut into sections ready for use.

PSL is very strong in its primary axis. The strength properties are higher than those of sawn timber. Additional strength is gained from the 10 % densification relative to the original timber density. Strands fail in tension only because the strand overlap is large and the resistance to shear is greater than the tensile strength of the strand.

2.3.2 OSB—Oriented Strand Boards

Oriented strand boards (OSB) are developed from wood strands which are typically 15–25 mm wide, 75–150 mm long and 0.3–0.7 mm thick, cut from logs of small diameter. Adhesive is used to bond the strands together and the boards are fabricated under pressure at a high temperature. The strands in the outside layers are aligned parallel to the length, whereas the internal strands are deposited randomly or perpendicular to the face layers. OSB is therefore a multi-layer board, used in a large variety of structures of an industrial or decorative nature. The boards are used for covering floors, ceilings and sometimes walls. The OSBs are also being utilized more and more in packaging, for cases, pallets, etc.

2.3.3 LSL—Laminated Scrimbed Lumber

Like PSL, laminated scrimbed lumber (LSL) uses a manufacturing technology which makes it possible to convert small trees into elements with a large cross section, but treetops and branches can also be used. The debarked wood is flattened and partially split in the longitudinal direction using crushing and coarse so-called scrimming rollers, and the dried "scrims" are further processed using fine scrimming rollers to form mats. These mats are in general layered to each other to achieve the thickness and properties required of the final component. The orientation of the mats in LSL is parallel (i.e. the fibre orientation of all the scrims run parallel to each other), perpendicular (i.e. different layers oriented at 90° to each other) and also mixed, before the mats are laminated together under pressure to a component of beam or panel size. In general a PF adhesive is used.

2.3.4 PB—Particleboards

Particleboards (PB) are manufactured from wood particles that have various dimensions, cut mechanically by chippers, flakers, etc., and are in general three-layered. The mechanical properties of the boards depend on the dimensions as well as on the orientation and arrangement of the wood particles used in the board manufacture.

The particleboard industry has been commercialized successfully throughout the world because of the favourable conditions for raw material supply as well as because of market demands. The PB process makes it possible to use wood trunks of small diameters and wood residues. Historically, the availability of raw materials was good and the cost was relatively low. However, this relationship has drastically changed in recent years as a result of competition from the need for biomass fuel.

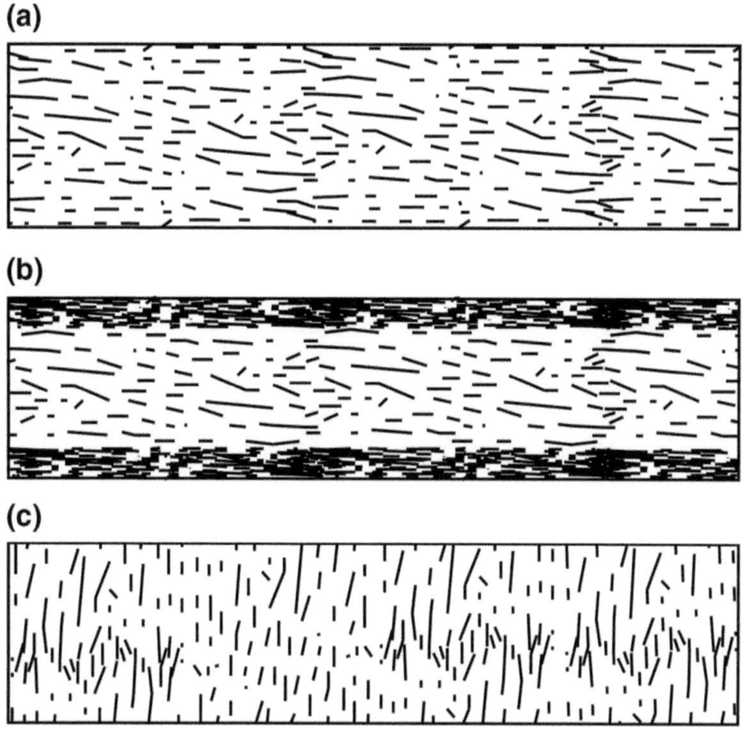

Fig. 7 Diagrammatic presentation of the cross section of particleboards, (**a**) manufactured by flat hot pressing, the orientation of the particles is parallel to the surfaces and the board is homogeneous (**b**) manufactured by flat pressing, board with three layers (**c**) manufactured by extrusion, the orientation of the particles is perpendicular to the board surfaces

Particleboards are very useful in the furnishing and construction industries. The production of particleboards also leads to an effective use of wood with a very small percentage of waste, 10–25 %, instead of 50 % in the sawing of logs.

The manufacture of particleboard is a dry process and there are two different methods of production; flat hot pressing and extrusion, which give different types of boards with different particle orientations. In the first method, the particles are oriented parallel to the panel surface, whereas in the second method the particles are oriented perpendicular to the surface. Figure 7 shows the two types of particleboard and the various stages in their production are illustrated in Fig. 8.

2.3.5 IBC—Inorganic-Bonded Composites

The traditional way of creating a wood composite is to blend wood together with an adhesive or a plastic as in WPC, and to press the blend at high heat and pressure. A considerably less used process that in which boards or other types of construction

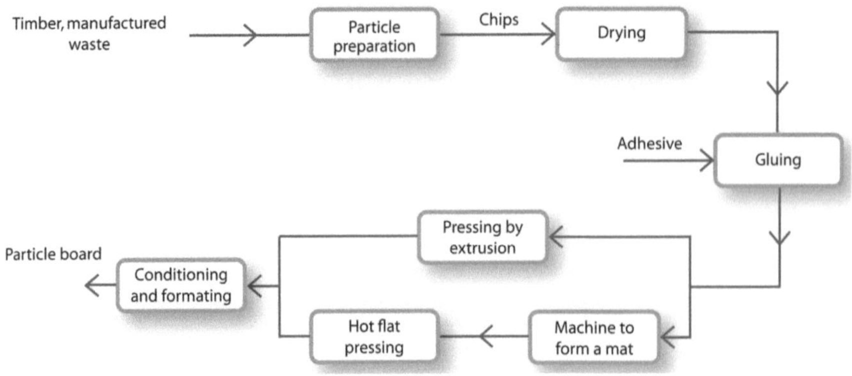

Fig. 8 Diagram of the various stages of the two different methods of manufacturing particleboards, i.e. extrusion and flat pressing

elements are produced from comminuted wood (excelsior, particles or flakes) or other types of fibre of vegetable biomass blended with cement, magnesite ($MgCO_3$) or gypsum. However, the most expedient binder giving strength, durability and acoustic insulation properties is Portland cement. In this process heat is not required.

Inorganic-bonded composites (IBC) have a long history that started with commercial production in 1914 in Austria that produced magnesium oxide-bonded wood excelsior boards (Maloney 1977; Geimer et al. 1994). Frybort et al. (2008) give a comprehensive review based on sources from the mid-twentieth century of the mechanical and physical properties of cement-bonded composites.

There are only about 50 mills worldwide making cement-bonded particleboard (CBPB), each of which produces on average only about 200 m^3/day. The wood–cement boards are used for specialized structural applications. They have outstanding properties in terms of their reaction to fire, durability, sound insulation and stiffness, which render the product most suitable for internal wall constructions in public places, the lining of lift shafts, the construction of cabling ducts, soffits, motorway acoustic fencing and the cladding of prefabricated house units. The cement-bonded board is harder and more resistant than its components alone, with a lower cost and lower density than concrete (Hein et al. 2009).

Chemical substances, especially extractives, present in the wood may delay or sometimes even impede the cement curing, making it necessary to use some process that increases the chemical compatibility between the wood and the cement (Savastano et al. 2000). Many manufacturers use additives like mica (silicate/phyllosilicate minerals), aluminium stearate and cenospheres (a light-weight, inert, hollow sphere made largely of silica and alumina and filled with air or inert gas) in order to achieve certain board qualities. Typical cement fibreboard is made of 40–60 weight % of cement, 20–30 % of fillers, 8–10 % of cellulose, 10–15 % of mica. Additives such aluminium stearate and PVA are normally used in quantities less than 1 %. Cenospheres are used only in low-density boards with quantities between 10 and 15 % (Kuroki et al. 1993).

2.3.6 WPC—Wood-Plastic Composites

Although EWPs based on strands, strips, chips and particles have been made with thermosetting adhesives for many years, only in the last three decades has a serious attempt been made to incorporate wood flour and chips into thermoplastic adhesives in order to produce wood-plastic composites (WPC). The term "wood-plastic composite" refers to any composite that contains wood particles and any of the thermosetting or thermoplastic polymers. In contrast to the wood-thermoset composites, wood-thermoplastic composites have seen a phenomenal growth in the United States in recent decades and for this reason they are often referred to simply as wood-plastic composites (WPC) with the understanding that the plastic is always a thermoplastic. New compounding techniques and interfacial treatments utilizing coupling agents make it feasible to disperse high volume fractions of hydrophilic wood in various plastics. These compounds can be continuously extruded, thermoformed, pressed and injection moulded into any shape and size, and they thus have offer the potential to replace natural wood in many applications. A WPC is in principal a composite of wood particles and a thermoplastic, with a dry weight percentage of the wood component typically in the range of 50–60 % (Klyosov 2007). Figure 9 shows an example of the use of a WPC product in Sweden. Today, WPCs are characterized as a building material and they have their main markets in the US (Clemons 2002). The European WPC market is also steadily increasing. In general, WPC products are marketed as a low maintenance building material with a high ability to resist fungal decay, although combinations of wood and polymers often have poor long-term durability when exposed outdoors. A major cause can be insufficient wood–polymer adhesion due especially to an intrinsically low compatibility between the wood substance and the polymers used. Adhesion losses are usually caused by the hygroscopicity of wood and the differences in hygrothermal properties between the components.

2.4 EWPs Based on Fibres

Wood fibres can be used to produce a wide variety of low-density three-dimensional webs, mats and fibre-moulded products. Short wood fibres can be blended with long fibres and formed into flexible fibre mats, utilizing physical entanglement, non-woven needling or thermoplastic fibre melt matrix technologies. The most common types of flexible mat are carded, air-laid, needle-punched and thermobonded.

A large number of EWPs based on fibres are developed in the form of panels. Their diversity is large and is increasing, taking advantage of scientific knowledge

Fig. 9 Example of extruded wood–plastic composite (WPC) profiles for cable channel covers (*dark coloured* WPC in the *rear* of the image) and end plugs for bobbins for large paper rolls (*front*). The WPC profiles are manufactured by Ofk Plast Inc. in Karlskoga and Polyplank Inc. Färjestaden, Sweden

and technological developments in the production field and taking into consideration the increasing requirements imposed on construction materials. The various types of fibreboards derived from wood are insulating fibreboards, and low-density fibreboards (LDF), medium-density fibreboards (MDF) and hardboards or high-density fibreboards (HDF).

The manufacturing processes are divided into two principal classes. In the first class, the cellulose fibres bind together by natural forces (hydrogen bonds). This bonding of the fibres is achieved through drying and compression in a process which takes place in a wet condition, no adhesive being used to bond the fibres together. This process is used to make paper, paperboard, insulating fibreboard, semi-hard boards and hardboards, Fig. 10. In the second class, the process is a dry process where the bonding together of the particles is accomplished by the use of various adhesives, the most widely used adhesives being PF and UF resins.

The panel material containing adhesive, i.e. MDF, is described in the following section.

2.4.1 MDF—Medium-Density Fibreboard

The essential difference between hard fibreboards and MDF is that adhesives are used as binder in MDF, whereas in hard fibreboards the lignin, under the effect of pressure and temperature is transformed into a kind of adhesive. The MDF manufacturing process began as a semi-dry process before being developed into fully dry process method. A diagram of the manufacture of MDF is given in Fig. 11. Since less water is used than in the wet process smaller amounts of polluted water are produced. In addition, this method allows the fabrication of panels with thicknesses from 2 up to 100 mm.

A uniform distribution of fibres during manufacture ensures that the MDF has a homogeneous structure, and it is possible to manufacture MDF boards with different characteristics to suit particular applications. The MDF is a homogeneous product with a density from 600 to 800 kg/m^3. A MDF board is easy to machine

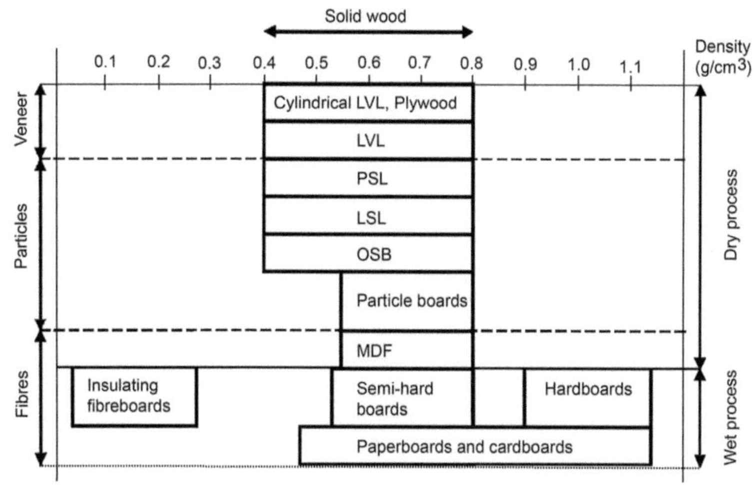

Fig. 10 Classification of panel materials according to the dimensions of the raw material and the manufacturing process (modified from Suchsland and Woodson 1986)

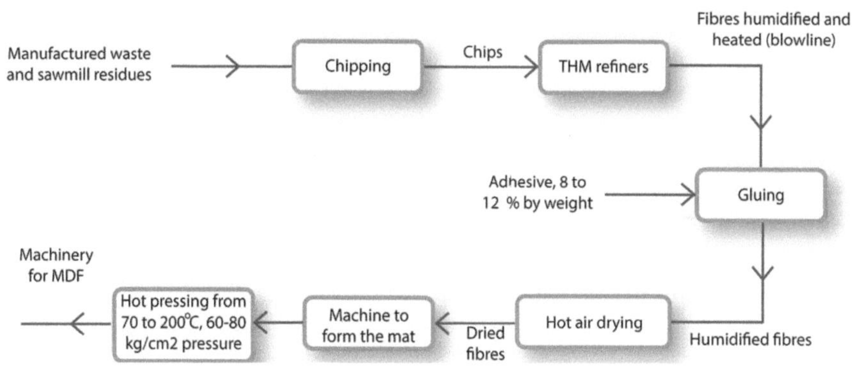

Fig. 11 Diagram of the manufacture of MDF by the dry process

and its regular surface is exceptionally well suited to painting or the application of a decorative coating. It is this quality which has given MDF the place that it occupies in the furniture industry. The thick MDF is used in joinery and for door frames, window frames, etc.

2.5 *Hybrid EWPs*

Hybrid engineered wood products consist of different EWPs or EWPs in combination with other materials. Typical hybrid EWPs are:

- Sandwich panels, e.g. a core of an insulation material, honeycomb, etc., and outer sheets of a board material such as plywood, OSB or MDF.
- Construction elements such I-joists or ribbed elements.

By combining different materials, the properties of the hybrid EWP can be designed specifically to give, e.g. low weight, high strength or low heat transmission.

2.6 Joining Members in Timber Engineering

In wood constructions, different kinds of semi-finished wood-based components are joined together to function as a system. Joints affect the structural behaviour of the elements, e.g. by providing stiff connections or hinges. Traditional joinery techniques have a very long tradition. A comprehensive overview of traditional Japanese, Chinese and European joinery techniques for sawn timber structure has been given by Zwerger (2012), and for constructions and joint techniques from logs by Phleps (1982). Joints in timber structures are in general classified as: traditional timber joints, glued joints and dowelled joints. Glued joints in timber engineering use the same types of adhesives as those used in the manufacture of structural elements like glulam, LVL and finger-jointing for structural purposes (see in the section on adhesives). A waterproof adhesive is strongly recommended as a fastener for wooden structures. Dowelled joints are the most common type of fastener for wood elements, and they transfer forces through shear in mechanical fasteners mounted at an angle to the force direction. The fasteners are made of a ductile steel material, and are designed for a large variety of different functions, Fig. 12. Metal fasteners and connectors can be divided into the following groups:

- Nails, screws, bolts and steel dowels
- Nail plates, punched metal plate fasteners, single- and double-sided tooth plate connectors
- Special connectors of metal plates, steel straps and framing anchors
- Steel rods for the cross-bracing of structures

Metal fasteners and connectors are easy to identify and distinguish from the wood material, and can be separated therefore reuse or before combustion of the recycled wood products (see Sect. 6 below).

The development of wood-to-wood joinery was not in the focus of research for many decades (Schindler 2009), due to the market-dominating position of metal fasteners. The craft sector was facing difficulties when transferring traditional knowledge about the joints load-bearing capacity in a suitable form for engineers and architects like calculation tables and assured load-bearing capacity. Metal fasteners were developed at the beginning of the twentieth century, and in the 1950s

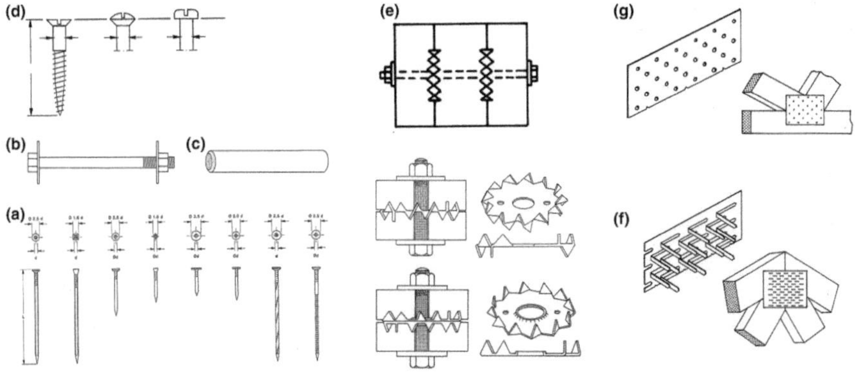

Fig. 12 A selection of timber fasteners and connectors made of metal: (**a**) nails, (**b**) bolt, (**c**) dowel, (**d**) screws, (**e**) single-sided toothed plate connector used for a demountable joint (*bottom*), double-sided toothed plate connector for a permanent joint (*middle*), and a joint with bolts and connectors (*top*), (**f**) punched metal plate, (**g**) nail plate for hand nailing

a variety of fastener types were offered and were continuously being used to replace traditional wood-to-wood joints (Graubner 1992). With the introduction of CNC technology in the carpenters workshop and the building industry, wood-to-wood joinery regained part of its former position and scientific attention, due to its economic and design potential (Schindler 2009).

One indication for this is the rediscovery of the dovetail joints in the 1990s for robotic manufacturing in Europe and North America. This rediscovery led to a focus on research on this type of joint, which finally resulted in technical approval, e.g. in Germany (DIBt 2012), and the great effort by, e.g. Bobacz (2002), Dietsch (2005), Hochstrate (2000), Holzner (1999), Kreuzinger and Spengler (1999), Tannert et al. (2007, 2011) which yielded a lot of understanding and established it for a multitude of applications. Figure 13 shows an example of a new concept for joining timber in construction by Rebstock et al. (2015). This so-called Makerjoint concept uses LVL as nodes in regions with a pronounced non-uniform stress distribution and sawn timber in regions with a more uniform stress distribution. No metal fasteners or adhesives are used in the joint between the timber and the LVL. The development of different wood-to-wood joints with robotic tools yielded designs which show the great potential of wood-to-wood joints. Since then different types of wood-to-wood joints have been used, e.g. in the Tamedia office building and the Sequential pavilion, both in Zürich, Switzerland and in the ICD/ITKE pavilion in Stuttgart, Germany.

Current development and research point towards an increased use of six-axis robots, which resemble a human arm and which were formerly most common in industrial manufacturing contexts. Here, Robeller with co-workers have recently developed snap-fit joints or dovetail joints for shell structures without glue (Robeller et al. 2014a, b).

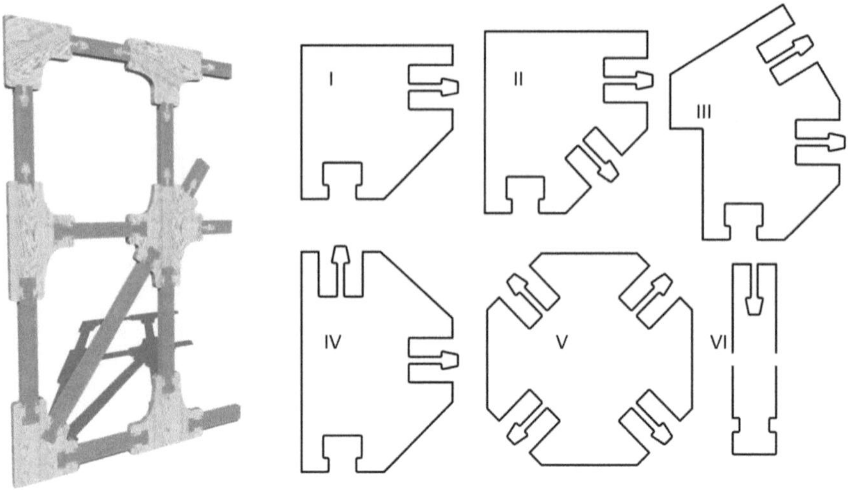

Fig. 13 The Makerjoint concept. (**a**) Sketch of a plane wall assembly, and (**b**) examples of nodes: I+II—corner nodes, III—roof node, IV—multiple floor node, V—bracing node, VI—Makerjoint beam (Rebstock et al. 2015)

2.7 Environmental Aspects of EWPs

Forestry and forest-related industries have never before been as focused as they are today in discussions regarding the major challenges of the future. Instead of utilizing the earth's limited resources, we have to use renewable materials, fossil fuels must be phased out and individual consumption must to a greater degree reflect the concerns for the climate and care for the environment. In this context, the emissions of carbon dioxide have been in focus for a long time. One way of reducing the emission of carbon dioxide is to use a greater proportion of timber products and to increase the life of these products so that the carbon is bound over a longer period of time. Another possibility is to replace energy-intensive materials with timber and timber-based products (Mahapatra and Gustavsson 2008). Timber and timber fibres have at least two or three utilization cycles before the material returns to the natural cycle. In the first cycle, the material is used in products such as timber, panelling, construction elements, furniture, etc. In the second cycle, the material can be used in a recycling process, i.e. as timber-based boards that are partly manufactured from recycled timber. Finally, in the third cycle, the timber material is used for energy conversion, e.g. in a power plant.

Kitek Kuzman and Kutnar (2014) have discussed the environmental impacts of primary wood products, and Werner and Richter (2007) reviewed the results of approximately 20 years of international research on the environmental impact of the life cycle of wood products used in the building sector compared to that of functionally equivalent products from other materials. Their study concluded that fossil fuel consumption, potential contributions to the greenhouse effect and quantities of solid waste tend to be much less for wood products than for competing products; impregnated wood products tend to be more critical than comparative products with respect to toxicological effects and/or photogenerated smog depending on the type of preservative; although composite wood products such as particleboard or fibreboard make use of a larger share of the wood of a tree than products made of solid wood, there is a high consumption of fossil energy associated with the production of fibres and particles/chips as well as in the production of adhesives, preservatives, etc.

Kutnar and Hill (2014) used a cradle-to-grave analysis to present the carbon footprint of 14 different primary wood products. The largest source of emissions for all sawn timber products is in removing the timber from the forest, while for kiln-dried sawn timber the drying process is a close second. For MDF and HDF, the extra energy required to convert the raw material to fibres, in addition to the energy required to apply pressure and heat to the products, is responsible for the bulk of the emissions from these products. The adhesives used in particleboard, plywood and OSB are responsible for the largest fraction of emissions from these products. This is especially significant considering the low total volume they represent in the final products. Glulam emissions derive mostly from the harvest and initial production of the softwood, but also from the extra energy required to apply pressure and set the adhesives used. Altering the system boundaries would yield different results. Furthermore, results would have been modified if the carbon footprint calculation took into account the carbon sequestration of wood, the use of recycled wood products and other similar issues pertinent to LCA. Furthermore, the results would have been different if a full life cycle of products, cradle-to-grave or cradle-to-cradle, had been considered.

There are many materials in the construction and housing industries that compete with timber, e.g. steel and concrete for frames and large constructions, bricks for walls and facades, and PVC and other plastics for windows, building features and furniture (Gustavsson et al. 2006). For timber as a material to be competitive against other materials, the timber's environmental advantages alone are not sufficient, i.e. that timber shows lower emissions of carbon dioxide according to calculations based on LCA criteria. Timber must also be competitive for its technical qualities, show a high material utilization during further processing, and not least, show a competitive economic yield during usage.

3 Adhesives

Bonding means connecting two solid bodies using an adhesive which fills the gap between the bodies and is able to transmit forces between the bodies. For wood, this means that the adhesive can penetrate beyond the surface and into cell lumens. An adhesive needs to be a liquid, or at least have sufficient flow under conditions of bonding so that it can come into good contact with the two surfaces to be bonded, and a good wetting and adhesion to the wood is always necessary. The adhesive must then solidify to hold the wood surfaces together, either by cooling and/or loss of water in the case of thermoplastics or by a chemical reaction to cross-link the material or increase the molecular weight in the case of thermoset adhesives. Chemical properties are important in bonding for developing adhesive and cohesive strength, while bond performance is measured as the mechanical strength for holding the wood surfaces together under various exposure conditions.

The art of gluing wood is very old. One of the earliest indications of the use of adhesive and veneer is found in Egyptian tombs and reliefs. Little is known of the methods used, e.g. how the wood was sawn into sheets or how the glue, which 3500 years later still holds thin layers of face wood to the heavier core, was prepared, but the work has endured and speaks for itself. Figure 14 shows a mural record found in Thebes that describes the production of an intarsia or plywood construction from about 1490–1436 B.C. Killen (2000) says that Egyptian carpenters began to laminate thin sheets of wood as early as the Third Dynasty (2686–2613 B.C.) in an attempt to fabricate a large sheet of wood which was dimensionally stable and equally strong in all directions. An example of six-ply wood, where the fibre orientation of one sheet is at right angles to the next, was discovered in a sarcophagus within the Third-Dynasty step pyramid complex of Djoser at Saqquara (Lauer 1933; Lucas 1936). There are also other examples of the early use of adhesives for the improvement of wood products. Insulander (1997) has studied the construction of the Scandinavian Saami two-wood bows. The oldest has been dated to ca. 200 B.C. but most of the finds are from the Middle Ages. This type of bow was constructed from two strips of wood, one of birch or other deciduous tree and one of compression wood from pine, which were glued together with an adhesive made from the skin from the perch fish (see also Schefferus 1673; Linné 1737). Protein-based adhesives were used until they were replaced by fossil-fuel-based systems beginning in the 1930s.

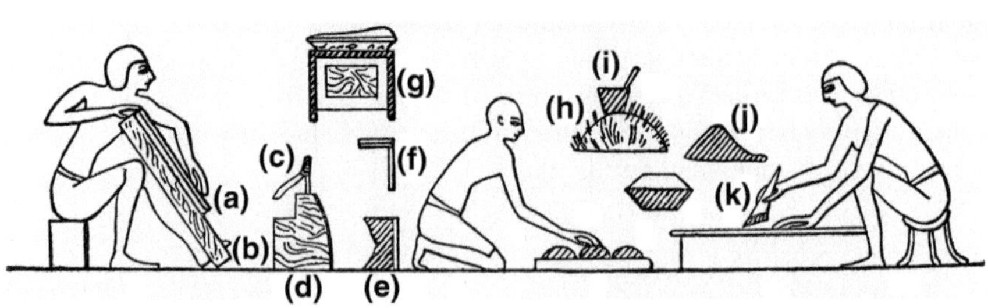

Fig. 14 Mural record of veneering, discovered in the Sculpture of Thebes and dated as early as the time of the Thutmosis III (1490–1436 B.C.). The man to the *left* is applying veneer on a core of ordinary wood, the man in the *middle* is grinding something and the man to the *right* is applying adhesive with a brush. The mural record also shows (**a**) a piece of wood (**b**) applied to one of ordinary quality and (**c**) an adze fixed into (**d**) a block of wood. Some tools and equipment: (**e**) a ruler, (**f**) a right angle, (**g**) a box, (**i**) a adhesive pot on (**h**) the fire, (**j**) a piece of adhesive and (**k**) a brush, from Knight and Wulpi (1927)

3.1 Different Types of Adhesives

Today, there are several wood-to-wood adhesives which produce joints stronger than the wood itself. Their use is, however, generally restricted to controlled wood-to-wood bonds that are well made, which require adequate, uniform pressure on smooth, clean, well-mated surfaces with even adhesive spread and factory conditions where temperature, adhesive age and formulation, press time, pressure and adequate wood moisture content can be carefully monitored.

It is difficult to determine with accuracy the use of adhesive in Europe. However, estimates suggest that the adhesive use for particleboard is split between UF (92 %), melamine UF (7 %) and isocyanates (1 %). UF is the primary adhesive used for MDF. OSB, on the other hand, is primarily made with polymeric diphenylmethane diisocyanates (75 %), while UF (10 %) and MUF (15 %) are also used (Kutnar and Burnard 2014).

Adhesives can be classified in many different ways. A division with respect to the chemical character and binding mode of the adhesive is probably the most useful. It is natural to begin with a breakdown between synthetic adhesives (see 3.2.1–3.2.9 below) and natural adhesives (see 3.2.10). The most common types of adhesives used in engineered wood are described below.

3.1.1 UF—Urea-Formaldehyde Adhesives

Urea-formaldehyde (UF) adhesives have several strong positive aspects: very low cost, non-flammable, very rapid cure rate and a light colour. UF adhesives are the largest class of amino resins, and are the predominate adhesives for interior products.

A major drawback of UF adhesives is their poor water resistance and they have high bond-line failure under accelerated ageing tests. Another area of concern is the long-term hydrolytic instability of these adhesive polymers, which generally show the least durability of any formaldehyde copolymer adhesive. UF adhesives are believed to depolymerize resulting in the continuing emission of formaldehyde. UF adhesives are typically used in the manufacture of products used in interior applications, primarily plywood, particleboard and MDF, because moisture exposure leads to a breakdown of the bond-forming reactions. Excessive heat exposure will also result in chemical breakdown of cured UF adhesives.

3.1.2 PF—Phenol-Formaldehyde Adhesives

PF polymers are the oldest class of synthetic adhesives, having been developed at the beginning of the twentieth century (Detlefsen 2002). These adhesives are widely used in both laminations and composites because of their outstanding durability, which derives from their good adhesion to wood, the high strength of the polymer and the excellent stability of the adhesive. PF adhesives are typically used in the manufacture of construction plywood and OSB where exposure to weather during construction is a concern. The PF adhesives can serve in almost all wood-bonding applications, as long as the adhesive in the assembly can be heated. In many cases, if moisture resistance is not needed, a lower cost UF adhesive can be used. Like most adhesives, the commercial products contain more than just the resin, depending on the application. The most common additive is urea to provide improved flow properties, to scavenge free formaldehyde and to reduce the cost.

3.1.3 MF—Melamine-Formaldehyde Adhesives

Like formaldehyde adhesives made with phenol and resorcinol, MF adhesives have a high water resistance, but they are much lighter in colour than the others. MF adhesives are most commonly used for exterior and semi-exterior plywood and particleboard, for finger-joints, for decorative laminates, paper treating and paper coating. MF resins are often used in combination with UF.

The limitation of the MF adhesives is their high cost due to the cost of the melamine. This has led to the use of MUF adhesives that have much of the water resistance of MF adhesives, but at a substantially lower cost. The MUF adhesives, depending on the melamine-to-urea ratio, can be considered as a less expensive MF that has lower durability or as a more expensive UF that has better water resistance (Dunky 2003). The MUF adhesives can replace other adhesives that are used for some exterior applications.

3.1.4 MDI—Methylene-Diphenyl-Diisocyanate Adhesives

Several classes of adhesives used in wood bonding involve the use of isocyanates, because of their reactivity with groups that contain reactive hydrogens, such as amine and alcohol groups, at room temperature. This allows great flexibility in the types of products produced, because they can self-polymerize or react with many other monomers. Isocyanates are most often used to produce PU by reacting with liquid diols.

Polymeric diphenylmethane diisocyanates (pMDI) are commonly used in wood bonding and are a mixture of the monomeric diphenylmethane diisocyanate and methylene-bridged polyaromatic polyisocyanates. The higher cost of the adhesive is offset by its fast reaction rate, its efficiency of use and its ability to adhere to difficult-to-bond surfaces. pMDI adhesives are used as an alternative to PF adhesives, primarily in composite products fabricated from strands, and they are sometimes used in core layers of strand-based composites, with a slower curing PF adhesive in the surface layers. The use of pMDI requires special precautionary protective measures because the uncured adhesive can result in chemical sensitization of persons exposed to it. A cured pMDI adhesive poses no recognized health concerns.

3.1.5 PU—Polyurethane Adhesives

PU are made up of long polyol chains that are tied together by shorter hard segments formed by diisocyanate and in some cases chain extenders (Pizzi and Mittal 2003). The polyol chains impart low-temperature flexibility and room-temperature elastomeric properties. The advantage of PU adhesive is the reaction of isocyanates with the active hydrogen on the surface, subsurface or air making possible the gluing of surfaces with different moisture contents. The curing time is relatively short, 3–4 h, resulting in a strong, water-durable bond. The disadvantages of the PU adhesives are isocyanate emissions and the higher pressure needed when clamping.

3.1.6 Epoxy Adhesives

Epoxy adhesives are two-component thermoset polymers based on an epoxy, epoxide or ethoxyline group resin and a hardener (Wallenberger and Bingham 2010). The advantages of epoxy adhesives are their good chemical and thermal resistance and low clamping pressure. The curing agent, the hardener, produces an insoluble, intractable, cross-linked thermoset polymer. The properties of the cured epoxy adhesive depend on the type of hardener and on the cure temperature.

Epoxy adhesives are currently mostly used for the fibreglass reinforcement of, e.g. wooden boats and glulam. Their advantages are stronger bonds with the wood, a high durability and a greater impact resistance than e.g. polyester adhesives, and they are relatively easy to work with. As disadvantages, epoxy adhesives are rather expensive, have long curing cycles. Also, most of the epoxy suffers from "amine blush", i.e. after application and during the curing process the epoxy releases a blush to the surface.

3.1.7 PVAc—Polyvinyl-Acetate Adhesive

Vinyl acetate homopolymers are simply made adhesive bases manufactured by addition polymerization in the presence of water and stabilizers (Pizzi and Mital 2003). External plasticizers, e.g. dibutyl phthalate are usually added to confer flexibility and to lower the temperature at which they form a film on drying (Pizzi and Mittal 2003). Higher quality products may be made by the copolymerization of ethylene with vinyl acetate to form ethyl vinyl acetate (EVA). The advantages of PVAc adhesives are their low costs, ease of use and minimum of harmful environmental effects. The disadvantages of PVAc adhesives are their low water durability. This drawback can be reduced by blending the PVAc with MUF resins.

3.1.8 Resorcinol Adhesives

Cold-setting adhesives with good water resistance are the resorcinol adhesives. Resorcinol-formaldehyde (RF) and phenol-resorcinol-formaldehyde (PRF) adhesives are mainly used in the manufacture of structural exterior grade joints. Among their net advantages of strong joints when setting at ambient temperatures, RF and PRF adhesive are rather unavailable and are thus expensive.

3.1.9 Polyester Adhesives

Unsaturated polyester resins consist of low molecular weight condensation products of unsaturated and saturated biacids and diols dissolved in a styrene monomer or other suitable reactive diluents (Wallenberger and Bingham 2010). The unsaturation of the adhesive systems provides vinyl sites for cross-linking and the creation of the thermoset adhesive. Cross-linking can be initiated by various activators at ambient or raised temperatures. Organic peroxides are used as initiators and cobalt complexes are used as accelerators, to reduce the minimum temperature at which the decomposition of the organic peroxide takes place. Polyesters are mainly used for the fibreglass reinforcement. The advantages of polyester adhesives are their wide choice, reduced costs and curing at room temperature. As disadvantages can be mentioned styrene emission and the rather low impact strength of the surface.

3.1.10 Bio-Based Adhesives

Bio-based adhesives were widely used prior to the early 1970s in construction plywood. In the mid-1970s, they were supplanted by PF adhesives, because of the superior bond durability provided by phenolics. The most common bio-based adhesives are protein-based, e.g. from animal bones and hides, milk (casein), blood, fish skins and soybeans. Natural protein adhesives are not useful at high moisture levels. Adhesives from renewable resources are, however, still of interest, not least for environmental reasons, but none had the performance and production cost to make an impact in the wood adhesive market until soy flour adhesives, using a polyamidoamine–epichlorohydrin (PAE) resin, were developed by Li et al. (2004). Other adhesive systems have been developed using soy proteins in combination with other adhesives. When soy flour becomes a part of the adhesive network rather than just being used as a filler, it can replace about half the phenol in basic or neutral formulations that meet the performance requirements for the face adhesives of, e.g. OSB (Frihart 2011). It is also possible to obtain bio-based adhesives from wood itself, e.g. tannin and lignin.

Tannins are used as a partial replacement of phenols because of their good reactivity, but the volumes are small because of their limited availability (Pizzi 2006). Tannins are more reactive than phenol but more expensive. Limitations of tannin-based adhesives, compared to synthetic adhesives, are their high viscosity, limited availability and inconsistent source and therefore reactivity. Tannin has been used as an adhesive in particleboard and MDF production, and for laminate and finger-joint bonding.

Lignin is available in large quantities at low cost, as a by-product of pulping processes for papermaking. Lignin is not as reactive as tannin with formaldehyde but it can be modified. Lignosulfonates from the sulfite pulping of wood have been found to be more useful for the production of reactive lignins (Rowell 2005). Lignin continues to be evaluated as a partial replacement of phenols, but the volume used in wood adhesives has been low because of its limited reactivity. New biorefinery technology may provide more reactive lignins in the future, and the amount of lignin that is available at a competitive price is substantial.

3.2 *Environmental Aspects of Adhesives*

Although wood is a natural material, bonded wood products have caused some environmental concern related to formaldehyde and other volatile compounds in the adhesive formulations used in bonded wood products. Heating increases the problem, as it raises the vapour pressure of reactive chemicals. Isocyanates can react rapidly with compounds in the human body. Both EPI and PU adhesives contain isocyanates. The preparation and processing of epoxy also poses significant health risks in the form of, e.g. allergies. There are also legal requirements which aim to depress the levels of emission especially of free formaldehyde.

Environmental interest and the rising cost of petroleum-based adhesives, i.e. synthetic adhesive, has made bio-based adhesive increasingly interesting. A recent example is that using polyamidoamine-epichlorohydrin (PAE) resin as a co-reactant has been found to be effective in increasing the wet bond strength of soy adhesives and has led to a resurgence in soy-based adhesive consumption (Frihart and Birkeland 2014).

The foremost area of environmental concern with regard to adhesives has been formaldehyde emissions from bonded products, mainly those using UF adhesives. Products bonded with UF-based adhesives such as plywood, MDF and particleboards are often used, e.g. in kitchen joinery and furniture and they may therefore lead to an increase in the level of formaldehyde in the indoor air. Formaldehyde can also be cleaved from acid-curing inks and varnishes. Formaldehyde can react with biological systems in reactions similar to those that are used for the curing of adhesives. The problem can arise from both unreacted and generated formaldehyde (Rowell 2005). Unreacted formaldehyde is a problem during the manufacturing operation and in freshly produced composites, but formaldehyde emissions from composites decrease with time after production. Formaldehyde can also be generated by the decomposition of some formaldehyde copolymer adhesives, in particular the UF adhesives. These adhesive bonds are more prone to hydrolysis, generating free formaldehyde.

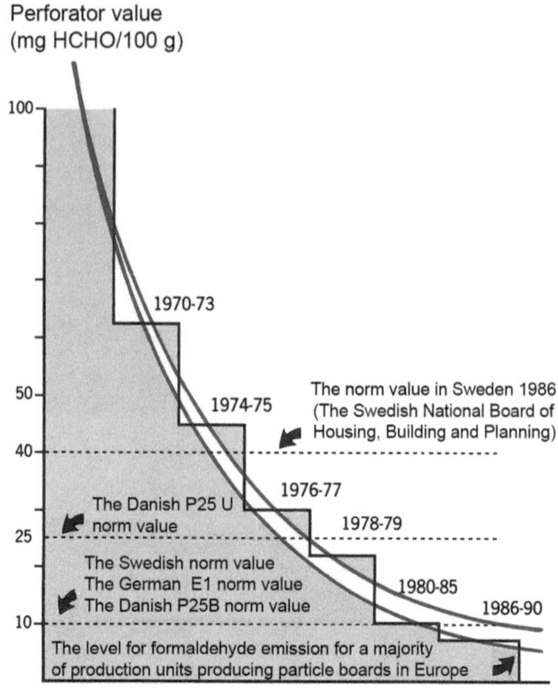

Fig. 15 Change in content of cleavable formaldehyde from particleboards in Europe during 1970–1990, measured as perforator value, expressed as mg formaldehyde per 100 g dry material. The perforator method is described in the European standard EN-120 (CEN 1992)

The greatest concern about formaldehyde is with particleboard, due to the large volume of indoor usage and the large amount of adhesive in the product. The particleboard industry has therefore together with adhesive manufacturers over a long period focused on reducing formaldehyde emission from the boards. Figure 15 shows the reduction in formaldehyde emission from particleboards in Europe from 1970 to 1990.

Richter (2001) provides a comparison of environmental assessment data of different wood adhesives. Little LCA data have been published so far for resins based on renewable resources or components (e.g. tannins, lignins, proteins). A study of the use of a lignin-based phenolic adhesive in combination with a laccase-initiating system has found a significant environmental impact associated with enzyme production (González-García et al. 2011).

4 Wood Modification

Strictly, wood modification involves a change in the macromolecular chemical composition of the wood cell wall, i.e. chemical or thermal modification. In this section "impregnation modification" where the cell wall is filled with an occluding substance is also included. Wood modification involves the action of a chemical, biological or physical agent upon the material, resulting in the desired property enhancement during the service life of the modified wood. The properties of wood are the result of the chemistry of its cell-wall components and the matrix they are in. If the chemistry is changed at the molecular level, the properties change and performance also changes. With the demands of society today, the modified wood should itself be non-toxic under service conditions and furthermore, there should be no release of any toxic substances during its life time, or at the end of its life following disposal or recycling. If the modification is intended for improved resistance to biological attack, then the mode of action should be non-biocidal. Historically, that has not been the case.

The market, especially in Europe, for new durable products of modified wood has increased substantially during the last few years. This increased interest depends partly on the restricted use of toxic preservatives due to an increased environmental concern. Another motive is the need for reduced maintenance. Three wood modification concepts, which have recently been commercialized, are acetylation, furfurylation and thermal treatment.

4.1 Preservation of Wood Against Biological Attack

Wood impregnation is in general a chemical treatment providing protection against various kinds of injurious organisms which decompose the wood in its use environment. The impregnation is carried out in industrial plants and is in many cases linked to the production line of the sawmill. Wood protection plants can historically be divided into impregnation plants, where the wood is treated to resist rot, and dipping plants, where the wood is treated to provide protection against mould and blue-stain fungi. The purpose of impregnation is to ensure that the wood, when it is used, shall have a longer life and resist attack from destructive organisms.

Dipping, on the other hand, does not give a product which is more resistant in use. It only prevents the wood from the blue-stain damage which can arise if the wood has too high a moisture content under unfavourable conditions, e.g. during storage before drying or in timber-yard drying. Dipping is not really an impregnation method, but from the 1940s it was frequently used in sawmills as a short-term protection against blue-stain fungus. Dipping has also been carried out at a few independent wood impregnation plants. The first dipping chemicals used were water-soluble fluoride-based agents. Later, phenolic agents (pentachlorphenol) came to completely dominate the dipping of sawn wood at the sawmills, but the need for dipping decreased when more and more sawmills started to instal kilns for wood drying instead of drying the wood outdoors, and in the 1970s the use of the toxic pentachlorphenol was forbidden as a dipping agent and the method is very seldom used today.

It is chiefly wood species with a low biological resistance which are impregnated, e.g. Scots pine, because the sapwood in pine, which has a low resistance in outdoor use, has open pores between the cells even after the wood has been dried, and this means that the impregnation liquid easily penetrates into the sapwood. Beech and spruce are also impregnated but not to the same extent as pine wood.

The birth of the wood impregnation industry dates back to the end of the 1850s. At first, only telegraph poles and railway sleepers were impregnated, and a copper sulphate solution (copper vitriol) was then used. When the first railways were built, it was realized that the sleepers must be protected against rot, and attempts were therefore made to impregnate sleepers with, e.g. copper sulphate solution according to the Boucherie method. Impregnation with creosote soon also became a relatively common impregnation method.

The Boliden mining company in Sweden, which started its activities in the 1920s, obtained large amounts of arsenic in the processing of iron ore, and in order to find a market for this by-product, research in the sphere of wood protection was initiated in 1932. Towards the end of the 1930s, sufficient knowledge had been obtained to make it possible to exploit the Boliden impregnation method to a greater extent. At this stage, the engineer Bror Häger was the R&D manager, and the Boliden impregnation salt developed by Häger contained arsenic, chromium and zinc. This agent was at first used in so-called open-tank plants. Pressure impregnation had thus come to stay. This meant considerable large-scale operation advantages compared with the Boucherie and open-tank methods and made possible an increase in the use of impregnated sawn wood, a product which in the 1940s became increasingly important. The development of new salt agents was also intensive during and just after the Second World War. Two products which came to have a great commercial importance and dominated the wood protection market in the north of Europe for more than two decades were Boliden K33 and KP Cuprinol.

The volume of impregnated sawn wood has increased greatly during the last few decades and today it exceeds the volume of impregnated sleepers and telegraph poles. There are several reasons for this: new spheres of application for impregnated wood were introduced (garden furniture, patios, playgrounds, etc.), the building of houses increased rapidly at the end of the 1960s and in the early 1970s, and there was a generally increased awareness of the advantages of using impregnated wood for certain purposes.

In the 1970s, impregnation of window frames with an oil-soluble protective agent according to the so-called double vacuum method was common. This type of impregnation has since then become increasingly widespread in the window joinery industry. Today, the wood impregnation activity at many sawmills is an important feature of their advanced refining activity. Substantial volumes of sawn wood, often of low quality, are impregnated in the spring for sale in the summer half of the year via building material retailers.

Most of the wood protection agents used are now classed as biocides and they are in many cases very toxic. The following impregnation methods have been used or are still being used today:

The French physician and chemist Auguste Boucherie (1801–1871) invented the Boucherie method for preserving wood by drenching it with mineral substances, usually copper vitriol, and this method was called "to boucherize" after him. In the Boucherie method, unbarked logs are injected with a protective agent (copper sulphate solution) through tubes at the root end. Since the tank is placed 10–12 metres higher than the logs, the protective agent is pressed into the log. After 5–15 days, the impregnation is seen to be complete when liquid begins to drip from the top end and via knots. The next stage is to debark the logs.

Open-tank impregnation is used principally for the impregnation of telegraph poles. The wood is submerged in a tub with an airtight lid. Steam is led into the tub and after 8–12 h, the wood is heated. After this heating, a cold impregnation solution is fed into the tub and the cooling leads to an underpressure, as a result of which the solution is sucked into the wood. This part of the process takes 30–35 h. The Boliden impregnation salt is usually used in this method.

There are a large number of pressure impregnation methods. The common processes are the Fullcell/Bethell method (ca. 600 l impregnation liquid/m^3 sapwood), the Rüping method (ca. 200 l/m^3) and the Lowry method (ca. 300 l/m^3). In the latter two impregnation methods, the uptake of impregnation liquid is less but, at the same time, the concentration of the active components is higher. The different methods give different penetration depths, which means that the products from the different methods have different spheres of application. In pressure impregnation, the main agents used are creosote and water-soluble metal-based agents.

The Fullcell method started to be developed in the middle of the nineteenth century and was based on impregnation with creosote, but since the 1940s chromium-based salts have been used. This is still the dominant process for salt impregnation. The process involves: the *pre-vacuum stage* (30 min) to removes air from the cell cavities in order to increase the liquid absorption, and the filling of impregnation liquid while maintaining the vacuum; *the pressure stage* (0.5–1 h) when the pressure is retained until no more impregnation liquid is absorbed, after which the liquid is evacuated; *the post-vacuum stage* (15–30 min) to make the surface of the wood dry for easier handling; *the draining and drying stage* to the desired moisture content.

The Lowry method was developed at the beginning of the twentieth century and was used chiefly on a small scale for chlorophenol-based agents in the 1960s and 1970s. The various stages of this process are: *filling* with impregnation liquid under atmospheric pressure, *the pressure period* and *the vacuum period.* During the post-vacuum period, excess liquid is pressed out and this makes the wood surface dry. For this reason, the concentration of impregnation liquid is twice as high.

The Rüping method was also developed at the beginning of the twentieth century. The impregnation agent used is creosote which must be heated to 100–120 °C before it is used.

Vacuum impregnation is used primarily for joinery materials for use above ground. Until the 1990s, substances having organic tin compounds as the active component were used, the different stages of the process being: *the pre-vacuum stage* to empty the wood cells of air; *the filling of impregnation liquid* while the vacuum is maintained; *the atmospheric pressure stage* when the cells are filled with impregnation liquid, *the pressure stage* to ensure that the impregnation liquid is spread as uniformly as possibly in the material; and *the post-vacuum stage* to ensure that the timber surface is dry.

Vacuum impregnation according to the Royal method was put into operation in 1970, primarily for slender materials such as panels and window timber. The process takes about four hours and is carried out in two stages at a temperature of about 85 °C; first a pressure impregnation with an ammoniacal copper agent and thereafter a vacuum period when the timber is treated with a special oil to reduce the uptake of moisture by the wood, sometimes with the addition of a coloured pigment.

Impregnating agents that are used or have been used can be divided as follows:

- Creosote (pressure impregnation),
- Water-soluble metal-based agents such as chromium-based salts and ammoniacal copper (pressure impregnation),
- Solvent-based agents (pressure impregnation),
- Solvent-based and water-soluble agents such as organic tin compounds (vacuum impregnation).

Impregnated wood is used in the same way as untreated wood and therefore bonding is also necessary after impregnation. Treated wood may be glued with synthetic adhesives of the phenol or resorcinol type, but the surface should be cleaned with steam, or wiped off in the case of oily preservatives. Gluing of treated wood requires a higher temperature and pressure. Wood treated with salt-type, waterborne preservatives is more difficult to glue and a higher temperature and higher pressure are needed.

4.2 Preservation of Wood Against Fire

Wood is a combustible material and this can sometimes limit its use. The wood's flammability can be influenced chemically with fire and flame retardants. Fire retardants provide protection against a fully developed fire and can affect the charring rate, ignitability, flame speed, smoke development and mechanical properties. Flame retardants are used for protection against the initial fire. The additives used in both cases are the same, i.e. mainly inorganic salts. Because fire-retardant treatment reduces the flammability of the wood product, fire-retardant treated wood products are often used for interior joinery in rooms, auditoriums, etc., where codes require materials with low surface flammability.

Fire retardant treatment of wood products, e.g. by chemical modification, may considerably improve their reaction to fire, and the highest fire classification for combustible products may be reached (Euroclass B). This allows a wider use of visible wood, both as interior wall and ceiling linings and as exterior cladding. Common problems of fire-retardant treatments are the risk of migration of the fire-retardant chemicals within the wood product and salt crystallization on the surface due to a high moisture content, and the loss of fire-retardant chemicals by leaching or other mechanisms. Both problems lead to a poorer fire performance, and also a risk of a change in the service life of the products. Some fire-retardant chemicals may cause corrosion on metals and affect the properties of the wood such as paintability, colour and mechanical strength. The toxicity of the fire retardants is an important factor that must also be considered.

The fire-retardant treatment may take place either by vacuum impregnation, by incorporation during the manufacturing of wood the products, or by surface treatment of the final product. In all cases, the amount of fire-retardant chemicals in the final product is decisive for its reaction to a fire. Normally 15–20 % of flame retardant based on the wood weight is added by impregnation, but large variations occur.

In vacuum impregnation, inorganic salts are the most commonly used fire retardants for interior wood products. These salts include ammonium sulphate, boric acid, diammonium, guanylurea and monoammonium phosphate, sodium tetraborate and zinc chloride.

There are a large number of commercial flame-retardant surface coatings on the market and it is not possible to describe them in detail. The coatings are divided into intumescent and non-intumescent. The most effective flame-retardant surface coatings are intumescent coating, as in a fire they develop insulating foam on the surface of the wood. Intumescent formulations include a dehydrating agent, a char former and a blowing agent. Potential dehydrating agents include polyammonium phosphate. Ingredients for the char former include starch, glucose and dipentaerythritol. Potential blowing agents include urea, melamine and chlorinated paraffins. 300–500 g/m^2 of paint is a common amount to give a good effect of a flame-retardant surface coating. Non-intumescent coatings include formulations of water-soluble salts such as borax, ammonium and diammonium phosphate.

4.3 Acetylated Wood

The idea of acetylation dates back to 1928 (Fuch 1928; Horn 1928; Suida and Titsh 1928) and later on pioneering work was performed by Stamm and Tarkow in the 1940s (Tarkow et al. 1946). Acetylation of wood is a so-called single-site reaction where one acetyl group replaces the hydrogen atom in one hydroxyl group in the wood cell wall and increases the naturally occurring acetyl content of wood from 1–3 to 20 % (Rowell et al. 1986). The acetylation process involves impregnation of the wood with acetic anhydride which is then reacted at elevated temperature, which results in esterification of the accessible hydroxyl groups in the cell wall with the formation as a by-product of acetic acid, Fig. 16. The resulting modified wood material has a lower equilibrium moisture content (EMC), a greater dimensional stability, maintained strength (but exhibits a brash failure mode), good weathering resistance under clear coatings, and superior resistance to biological degradation. The acetic acid must be removed from the wood material, otherwise the product will smell of acetic acid.

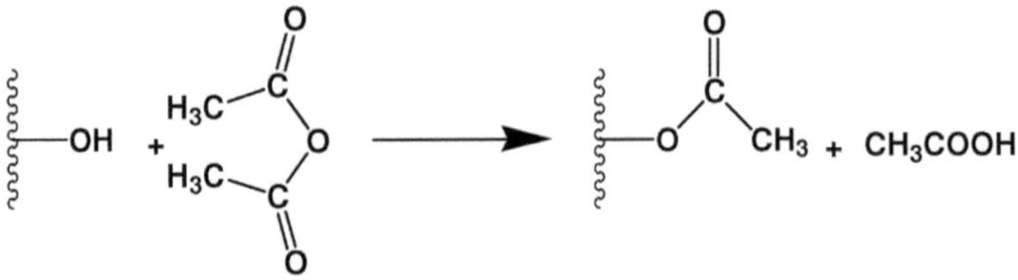

Fig. 16 The reaction of acetic anhydride with hydroxyl groups (−OH) in the cell wall in wood resulting in acetic acid as a by-product (CH_3COOH)

4.4 Furfurylated Wood

The wood furfurylation process involves pressure impregnation of wood with furfuryl alcohol, which is polymerized and reacted within the cell wall at elevated temperatures. The resulting material has a high dimensional stability, improved mechanical behaviour, except for impact resistance, and improved resistance to fungal decay. The research concerning furfurylation was also pioneered by Stamm and co-workers in the early 1950s (Goldstein 1955) and the commercial wood furfurylation process and characterization of the properties is described by Lande et al. (2004).

4.5 Thermal Modification

In recent decades, developments in the area of thermal treatment have accelerated considerably. During the 1980s, French and Japanese industries began to modify wood with the help of heat to increase the resistance to microbial attack. Since then, interest in thermal treatment has increased and several thermal treatment processes for sawn timber have been developed and industrialized in Europe.

Thermally modified timber (TMT) is, according to CEN (2007), wood in which the composition of the cell wall material and its physical properties have been modified by exposure to a temperature higher than 160 °C under conditions of low oxygen availability. The wood is altered in such a way that at least some of the wood properties are permanently affected. This product is related to heat-treated wood, but to distinguish it from heat sterilization at lower temperature (≈55 °C) with the purpose of killing pests in solid wood materials and preventing their transfer between continents and regions, we use the terms thermal treatment/processing and TMT. Thermal treatment significantly influences the properties of wood, e.g. hygroscopicity and dimensional stability, resistance against fungi and insects, mechanical properties, and also properties such as colour, odour, gluability and coating performance, Table 3.

In the beginning of the twentieth century, the use of heat and moisture in wood processing came into focus. It had been observed that wood dried at a high temperature changes colour, has a greater dimensional stability and has a lower

Table 3 Main differences in properties of thermal-modified timber (TMT) compared with untreated wood

Desirable property changes	Undesirable property changes
Lower equilibrium moisture content	Lower modulus of rupture (MOR) and to some extent lower modulus of elasticity (MOE)
Greater dimensional stability	Lower impact strength
Greater durability against decay	Greater brittleness (complicates e.g. machining)
Lower thermal conductivity	Lower hardness (Brinell hardness)
Lower density	
Dark brown colour	
Characteristic smell	
Longer pressing time for gluing	

hygroscopicity (Tiemann 1915; Koehler and Pillow 1925). After the First World War, comprehensive studies were made on the effect of the kiln drying temperature on the strength of wood for the aviation industry in the United States (Wilsson 1920). Stamm et al. (1946) reported the first systematic studies on wood thermal treatment, illustrating an increase in dimensional stability and an increase in the decay resistance of wood treated at a temperature between 120 and 320 °C, but their results were not really industrialized until the mid-1990s when several thermal treatment processes were commercialized.

Most industrialized thermal treatment processes today involve temperatures between 150 and 260 °C for times of several hours. The thermal treatment of wood above 300 °C is of limited practical value due to the severe degradation of the wood material (Navi and Sandberg 2012). The goal when treating wood is to increase its dimensional stability and resistance to biological degradation. The thermal treatment process is in most cases performed in vacuum, in air or in an inert gas such as nitrogen. Preheated oil can also be used, in which case the oil acts as a heat transfer medium and also excludes oxygen from the wood.

During thermal treatment, the chemical structure of the wood is transformed by autocatalytic reactions of the cell wall constituents. It is known that during the thermal treatment of wood under moist conditions, carbonic acids, mainly acetic acid, are initially formed as a result of cleavage of the acetyl groups particularly of hemicelluloses (Dietrichs et al. 1978; Bourgois and Guyonnet 1988). Depending on the acid concentration and on the temperature, hemicelluloses are hydrolysed into oligomeric and monomeric structures (Klauditz and Stegmann 1955; Carrasco and Roy 1992). The weight loss from hardwoods is greater than that from softwoods, probably due to the greater content of acetyl groups in hardwoods (Hillis 1975). Subsequently, the monomeric sugar units are dehydrated to aldehydes, furfural being formed from pentoses and hydroxymethylfurfural from the dehydration of hexose sugar units (Ellis and Paszner 1994).

The only type of thermal treatment process that adds a substantial proportion of external substances to the wood material during processing is that which using oil as process medium. The oil can be a plant oil such as rapeseed oil, sunflower oil, soybean oil. Tall oil or black liquor derivatives, in addition to drying oils such as linseed oil are also conceivable. Linseed oil proved to be unproblematic, although the smell that develops during the thermal treatment may be a drawback.

Thermal-treated woods are characterized by considerable changes in their chemical properties, resulting in a significant odour (Kamdem et al. 2000). The degradation products of a range of wood components cause this typical odour (Boonstra et al. 2006). The formation of toxic polyaromatic compounds during thermal treatment has been discussed in the literature. It has been suggested that the formation of these products leads to increased resistance fungal decomposition and micro-organisms (Kamdem et al. 2000). Increasing indoor applications of thermal-treated wood have led to an interest in elucidating the chemical composition of these emissions and in developing ways to reduce them.

Thermal treatment of wood is an innovative process currently being implemented in industrial applications. Although many technical aspects of thermal treatment are well known, the fundamental influence of the process on product performance, the environment and end-of-life scenarios remains unknown.

5 Surface Treatment

A coating is a covering that is applied to the surface of an object, usually referred to as the substrate. Paints and lacquers or varnishes are coatings that usual have dual purposes; to protect the substrate and to be decorative. In the case of wood, the coating is both decorative and functional. The surface should have an attractive colour or exaggerate the natural appearance of the wood, and at the same time protect the surface or the function of the wood product.

The largest segment of the paint industry produces architectural coatings, including consumer paints, which make up approximately half of the total quantity of coatings produced annually in the world. These products are used to beautify and maintain the surfaces in different places in society. Coatings applied at the time of manufacture of a product are known as industrial coatings, and they represent about 50 % of all the coatings used worldwide.

5.1 The Components of a Coating

The function and properties of a coating depend on how well the substrate, in our case wood, interacts with the coating. The properties of a coating are strongly influenced by its main components which are:

- Binders
- Pigments
- Solvents
- Fillers
- Additives

The binder binds the pigment particles to each other and to the wood as a protective layer, and it features a range of properties such as wetting, film formation and penetration that can be altered and controlled. Organic binders can be described as drying adhesives. Today, most organic binders are synthesized from crude oil, but there are still conventional binders based on oils from plants such as linseed oil. Typical binders include nitrocellulose, alkyd resins, amino resins, acrylates and polyester resins.

Pigments give colour and contribute to making the surface layer opaque. The pigment grains are usually so small that they cannot be distinguished by the naked eye and can also contribute to other properties of the coating such as diffusivity and gloss. Both soluble and dispersed pigments are used for stains.

With added solvents, the coating has a viscosity so that it can be applied to the wood. The solvent may also affect the penetration of the coating into the wood. Common solvents are water, white spirit, alcohols and turpentine. Earlier, xylene was a common solvent, but because of its high toxicity it is no longer used.

Fillers are used to modify the colour strength, the gloss of the coating and the extent to which the coating covers the substrate. Coating manufacturers also use fillers to reduce the manufacturing cost for the coatings.

Additives in coatings may include fungicides, alcohols or glycerol, to prevent mould growth, make the colour drip-free, accelerate the drying or improve adhesion, and to regulate certain properties of the finish such as gloss, consistency, wetting, flow, blister prevention and sandability. Specifically, polymeric coatings need a high amount of additives that provide desirable features such as wettability, preservative properties and cushioning foaming.

Coatings are available in a very wide variety depending on the properties, use, price, etc., and can be roughly divided into water-soluble and water-free coatings, Fig. 17. The water-soluble coatings can in turn be subdivided into two groups according to how the binder, pigment and solvent interact: slurries and dispersions. In a slurry the binder and the pigment are suspended in the solvent. If the coating is left, the pigment settles to the bottom and stirring is necessary before each new application. This separation does not occur in a dispersion, in which the pigments are so small that they remain suspended and the coating constitutes a suspension. In emulsion coatings, comminuted oil is dissolved in water with the help of an emulsifier. All water-soluble paints dry by evaporation of the water.

Another type of division is into organic and inorganic coatings, referred to as pigments. Organic colours are based on carbon compounds. Typical water-soluble organic coatings are those containing organic pigments, the inorganic pigments usually being classified as white, black and coloured pigments, Table 4.

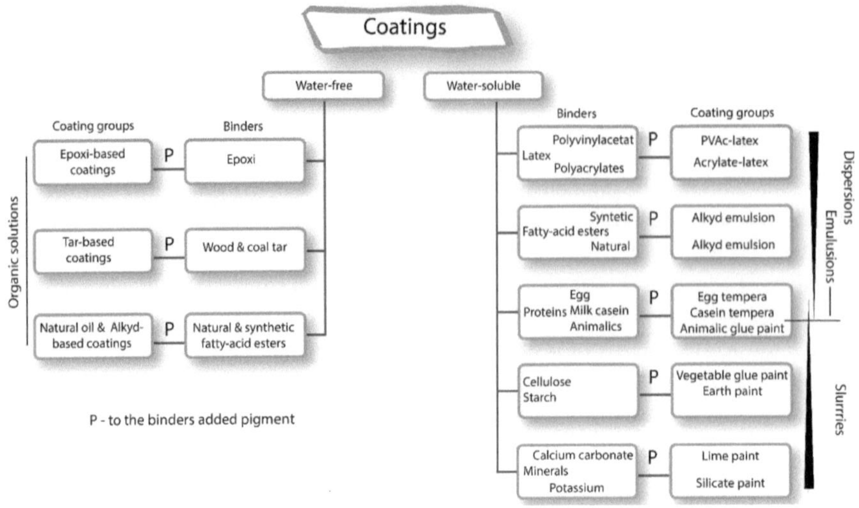

Fig. 17 Classification of water-free and water-soluble coatings based on binders

Table 4 Examples of organic and inorganic pigments commonly used in coatings

Organic pigment	Inorganic pigment			
	White	Black	Coloured	
Aniline (Phenylamine, $C_6H_5NH_2$)	Titanium white	Carbon black	Earth pigments (ochres, umbras, Terra di siena…)	Green earth
Azo compound (derivatives of diazene, R–N=N–R′)	Zinc white	Ferric black	Ferric yellow	Cobalt green
Phthalocyanine	Chalk	Bone black	English red Swedish red	Silica and silicates
Petroleum	Dolomite	Coal	Red lead	Ferrycyanide
Coal		Ferric oxide	Ultramarine Prussian blue	Chrome oxide

Depending on the binders they contain, lacquers, paints and fillers pass through different phases during the curing and drying process, and may therefore be classified as water-soluble coatings, air-drying coatings and reaction-curing coatings.

5.2 *Water-Soluble Coatings*

Water-based coatings are mainly thermoplastic systems, acrylate dispersions being the most common. These can be one-component, two-component or UV-curing systems, and in terms of quality they have improved significantly. There is now a water-soluble alternative for every application area and these products cure faster than the more conventional coating systems. Environmental legislation concerning reduced solvent emissions are forcing industries to increase the use of water-soluble systems or to invest in various forms of emission control and cleaning systems.

5.3 *Air-Drying Coatings*

Air-drying coatings dry through evaporation of the solvent/water. These include cellulose-based lacquers, where the drying time can be sharply reduced by heating and good ventilation.

5.4 *Reaction-Curing Coatings*

In this group of coatings, the curing process is triggered by a chemical reaction. In many cases, reaction-curing finishes are two- or multi-component products.

5.4.1 Acid-Curing Systems

Acid-curing systems are based mainly on alkyd and amino (urea or melamine) resins, often combined with nitrocellulose. The acid component, the hardener, is a catalyst that starts and maintains the binder's curing process until the chemical reaction is complete. The solvent evaporates prior to and during the chemical reaction, and plays no part in the final film. The curing of acid-curing coatings can be dramatically accelerated by applying heat. In most respects, acid-curing coatings offer the optimal combination of required properties, i.e. they combine a competitive price per unit area with high durability, rapid throughput and an uncomplicated production process. They can also be combined with "mild" solvents and have a relatively high dry solids content. A disadvantage is a limited release of formaldehyde in connection with the curing process.

5.4.2 Unsaturated Polyester Coatings

In unsaturated polyester coatings, a cobalt catalyst and peroxide hardener are added to initiate and maintain the curing reaction. In this case, a solvent must also be added. This may be styrene that also functions as a reactive element in the coating system and is included in the final coating, or a conventional solvent that must first evaporate. Polyester resins offer several benefits, including a higher dry solids content combined with high reactivity. The disadvantages are a limited pot life and the fact that both hardener and catalyst require careful storage and handling. The pot life can be extended considerably by using mixing pumps that feature multi-component dosing, or curtain coaters with two wet-on-wet heads. In this case, the catalyst is mixed into the lacquer from the first head, while the hardener is mixed from the second.

5.4.3 Polyurethane Coatings

Polyurethane (PU) coatings are two-pack systems in which hydroxyl groups (the binder) react with isocyanate resin (the hardener). With pure PU coatings, applying heat does not appreciably accelerate the curing process. With blocked PU products, however, heat does accelerate the curing process. These types of product have been attracting renewed interest in many markets because of regulations covering formaldehyde-free processes. They offer a promising alternative to acid-curing lacquers, thanks to their rapid development in recent years, in combination with a sharp reduction in the previously high level of free isocyanate.

5.4.4 UV-Curing Coatings

UV lacquers are products that are cured by exposure to ultraviolet (UV) radiation in the wave length of 200–420 nm. They may be based on unsaturated polyesters, although acrylate-modified polyesters, PU or epoxy binders, are more common. These are also called pre-polymers. Polyester is often diluted with styrene, vinyl ether or an organic solvent. Acrylate pre-polymer is often diluted with a low viscosity binder. These are not volatile like an organic solvent, but form a film by cross-linking with the pre-polymer. A photoinitiator must be present to catalyse the cross-linking process. Its purpose is to transform UV radiation and initiate the curing process by means of a chain reaction. This curing process is very rapid. UV-curing products with a dry solids content of 100 % are best suited to roller coating. Single-component pigmented UV-curing lacquers require UV lamps, known as "Ga-lamps", featuring a staggered wavelength area (400–420 nm). Otherwise, normal UV lamps (200–400 nm), known as "Hg-lamps", may be used. Designing a pigmented UV system for roller coating demands a thorough knowledge of the roller coater technique to achieve a superior finish. Combined systems with water-based primers and UV-curing products may be a more economically viable solution for some surface grades.

5.4.5 Water-Based and UV-Curing Coatings

Water-based UV-curing coatings can be one- or two-component systems, for clear or pigmented surface treatments. One-pack systems are the most common and are suitable in most situations and can replace most solvent-based systems (e.g. acid-curing or PU systems). To ensure proper curing of the edges when a one-pack system is used, UV ovens with inclined lamps must be used. The advantage of such a system, apart from the fact that it can be curtain coated or sprayed, is that once all the moisture has been removed and the UV curing has been completed, it produces a non-thermoplastic finish that is fully comparable with the high durability commonly associated with a conventional UV-curing system. Two-pack systems are used when extra high demands are placed on the finish or when inclined lamps are not available.

5.5 *Stains*

Stains are used to add colour to wood while revealing the texture of the substrate. They consist of colour and very finely dispersed transparent pigment, water, solvent and binder. Traditional stains mainly involved solvent-based stains, but in recent years water-based stains have been developed and they now dominate the market. Two factors in particular have contributed to this. Water-based stains make the texture of the wood surface appear more consistent and there are obvious environmental gains.

5.6 *Wood Surface Treatment with Oils and Waxes*

5.6.1 Oils

Oil has a strong, penetrating effect on wood, varying according to the structure of the substrate. How the wood has been prepared has a dramatic effect on the amount of oil the wood absorbs. A solid but unevenly grown piece of wood will exhibit very different absorption characteristics and may give the oiled surface a blotchy appearance. Normally, the surface must be dried and perhaps sanded again before the next layer of oil can be applied. Oil can be lightly tinted to function both as a stain and surface protection. In this case, a smoothly sanded surface with consistent absorption characteristics is critical. Add no more than 2–3 % of tinting paste, to avoid smearing. Solvent-free oils are ideal for the industrial finishing of pine and other wood species. All oiled surfaces require periodic aftertreatment to retain their protective properties.

Oils used for the treatment of wood are generally mixtures of various oils, and they may include alkyd, resin, balsam and solvents, formulated to give the product specific properties with respect to its use and/or price. Oils for the surface treatment of wood come mainly from the plant world, but animal oils and mineral oils also occur.

5.6.2 Waxes

Waxes are normally classified as “cold waxes” and “hot waxes”. Cold waxes are usually water-based but they are not always completely solvent-free. They are also colourless, but can be tinted to provide an attractive "stain-like" appearance. An opaque finish can be achieved by multi-layer waxing, as can special effects (metallic, etc.) and plain finishes. Hot wax finishes require a special type of application equipment and they generally provide better protection than cold waxing. Hot wax can also be tinted but this involves a much more complex process, as a result of which the entire hot wax container acquires the same tint, making it very difficult to clean. Even after hot waxing, the product still requires regular care.

5.7 New Surface Treatment Systems with Nanotechnology

The development of new surface treatments for wooden materials for, e.g. outdoor use is a continuous process. One region in which intensive research and development activities are in progress is that of so-called nanotechnology-based surface treatment. Nanotechnology is an innovative generic field of technology which is rapidly developing, where nanoscience can be briefly described as the study and understanding of the properties of materials and underlying phenomena at the nanometre level ($1\text{–}100 \times 10^{-9}$ m). Within nanotechnology, nanomaterials are being developed with completely new properties compared with those of traditional materials. It is a question of building up structures with special functional properties at the nanometre level and of utilizing these for the development of new materials, new components and new systems. Applied nanotechnology appears to have an extremely large development potential and most progress has so far been made in the fields of material technology and medical applications.

Many surface treatment products based on nanotechnology are now commercially available with an ability to produce surfaces which repel dirt and water and prevent the growth of algae, fungi and mosses, with an improved UV- and temperature-resistance, better colour permanence, greater scratch- and abrasion resistance, anti-graffiti properties, and insulating properties and which are even more environment-friendly than the traditional surface treatment products. If all these new nanotechnology-based products live up to their marketing promises, they will in the not too distant future lead to a revolution in the surface treatment of, e.g. external wooden façades.

Most of the products are, however, new and in many cases there are still unanswered questions relating to their long-term performance and technical lifetime, their maintenance and thus the total economy seen in a life cycle perspective for the product or system of which the surface treatment is only a small part. Most nanotechnology-based surface treatment systems currently on the market are intended for mineral substrates and few of them can be recommended for the surface treatment of outdoor wood. Nevertheless, a lot of development work is in progress and it is believed that it will be possible to modify many of the surface treatment products based on nanotechnology which are currently being used on mineral substrates so that they can also be used on outdoor wood.

5.7.1 Nanotechnology for the Surface Treatment of Wood

Surface treatments based on nanotechnology are currently available primarily for the treatment of substrates of an inorganic nature, such as metals, concrete, plaster, etc., which have the common feature that movements due to moisture or temperature variations are small. With wood as the substrate, the conditions and demands for a surface treatment system to function in a satisfactory manner are however different. This is probably the reason why it is more difficult to develop surface treatment products intended for wood. In the case of wood, the focus has been on surface treatment for use in an outdoor environment, rather than on, e.g. for house façades, for which it is more difficult to produce robust surface treatment systems.

In principle, nanotechnology offers an opportunity to produce composite materials with properties which can unite the special characteristics and advantages of different traditional materials. It is possible for instance to combine hardness with elasticity, good moisture protection with a high water vapour permeability, and good adhesion to the substrate with an ability to repel dirt and prevent microbial growth on the surface. With nanotechnology, it is also possible to achieve a surface coating to better protect the substrate against both climate-related and chemical environmental factors. Properties of a wooden façade which can be achieved by utilizing nanotechnology are:

- Surfaces which repel dirt and water
- Surfaces which hinder the growth of algae, fungi and mosses
- Surfaces which reduce moisture transport through end-grain sections
- Surfaces with a stronger resistance to UV radiation and temperature
- Surfaces with a better colour fastness
- Surfaces with a better resistance to scratching and abrasion
- Surfaces with more environment-friendly properties
- Surfaces which are anti-graffiti-treated
- Surfaces which reduce heat losses

The products available to make a surface dirt repellent or to hinder or prevent the growth of mould, fungi and algae can function according to several different "nanotechnological" principles. One principle is a treatment which makes the surface strongly water-repellent or "superhydrophobic", so that any water on the surface exists in the form of almost completely spherical drops. Particles of dirt or water-soluble contaminants can thus be transported away from the surface when the water drops run-off the surface. Another principle is to employ a treatment which makes the surface strongly hydrophilic at the same time as the treatment itself acts as a barrier to prevent water from penetrating into the substrate. The advantage in this case is that any water is retained as a thin homogeneous film which can run off the surface and dry without leaving any residual contamination. The third principle is to use photochemically active nanoparticles in the surface coating, e.g. a photochemically active form of titanium dioxide, which through photocatalysis can break down organic contaminants on the surface. The products resulting from the degradation by photocatalysis of contaminants on the surface are washed off by rain. A disadvantage of using photochemically active titanium dioxide particles for decontamination is however that if the binder contains organic components these will also be degraded by photocatalysis. This may make the method unsuitable for the surface treatment of wood.

If the surface of, e.g. a wooden façade could be made dirt repellent and able to prevent undesirable growth, the maintenance costs would be lowered be reducing or perhaps completely eliminating the need for façade cleansing. This property is already available for window glass.

Nanotechnology also offers the possibility of achieving a surface coating which can protect the wood substrate from ageing to a greater extent than traditional paints. In most of the information material relating to surface treatment products based on nanotechnology which are currently available on the market for use on outdoor wood, better weatherproofness and better colour fastness are mentioned as clear advantages over the traditional paint systems.

5.7.2 Paint Systems on the Market Based on Nanotechnology

The paint systems based on nanotechnology which are currently available on the market are here briefly described. Several of the systems are based on the so-called lotus effect, i.e. the self-cleaning mechanism found on the leaves of the lotus flower, the mechanism of which was first described and developed by the botanists Wilhelm Barthlott and Christoph Neinhuis at Bonn University in Germany (Barthlott and Ehler 1977; Barthlott and Neinhuis 1997; Neinhuis and Barthlott 1997). Superhydrophobic surfaces are extremely difficult to wet and the contact angle of water on such a surface is greater than 150° and the run-off angle is less than 10°.

Any dirt on the leaves of the Lotus flower rapidly disappears when it rains. The water drops roll over the leaf, apparently without any friction and remove at the same time any dirt particles which have collected on the surface of the leaf. Electron micrographs show that the surface of the lotus leaf has a mixed micro/nanostructure built up of waxed papillae covered with hair-like nano-sized tubes, Fig. 18. This double structure captures air beneath the rain drops which fall onto the leaf and creates a surface which effectively repels the water. A wax is secreted as a thin layer on the papillae which in turn lift up the water drops so that the dirt runs off the leaf. The adhesion between the dirt particles and the leaf is extremely low. Experimental studies of the lotus effect have focussed mainly on how drops with a diameter of the order of a millimetre behave when they contact the surface of the leaf.

The lotus concept was introduced when Wilhelm Barthelot entered into a cooperation with the company Sto Corp., as a result of which a Bionic[1] technology was developed to produce a synthetic surface with properties similar to those of the

[1]The word *bionic* is based on the Greek βίον, *bion*, which means life. Bionics is the application of systems found in nature to modern technology.

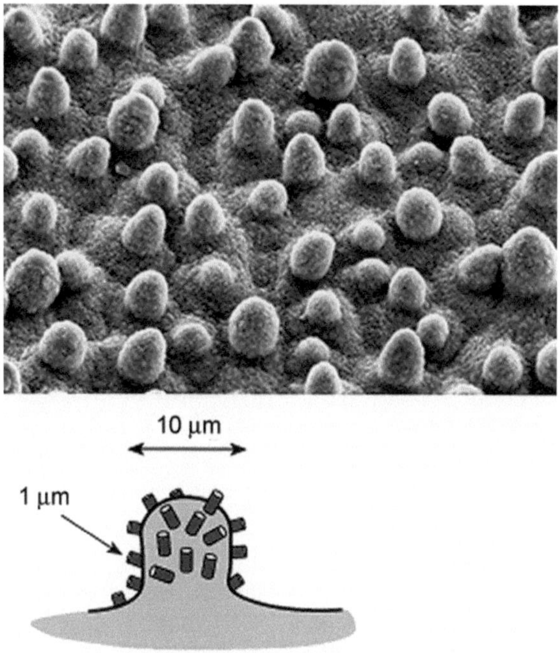

Fig. 18 The surface of the leaf of the lotus flower consists of small papillae covered with hair-like tubes

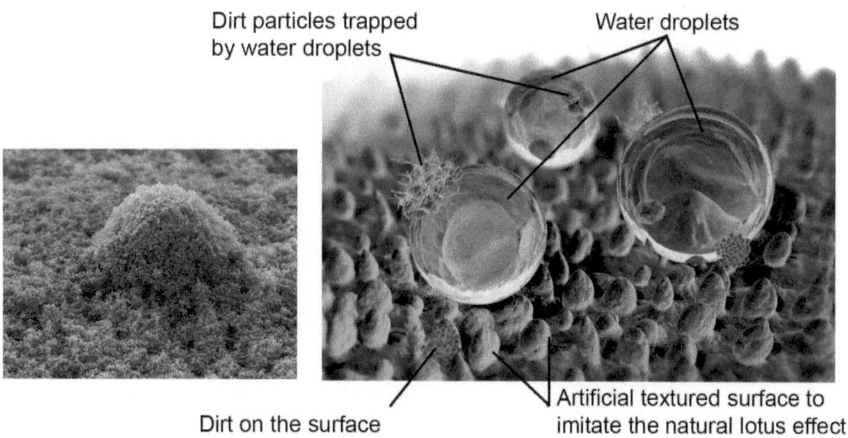

Fig. 19 *Left* Micrograph of a surface treated with a self-cleaning paint (ca. 10,000× magnification). *Right* Illustration of the mechanism for self-cleaning according to the lotus effect

lotus leaf and thus a surface which was dirt repellent, Fig. 19. The paint, which was introduced onto the market in 1999, is intended primarily for façades with a mineral substrate. There are a large number of references showing that the paint functions as intended all over the world. For the painting of outdoor wood, StoCoat Lotusan 215, 216 and 216D have been developed and they are recommended for use on wood which has been treated with a primer. The product is said to give a surface which is extremely hydrophobic, has an extraordinary ability to reject dirt and an improved resistance to moulds and algae. The paint is water-based and permeable to water vapour with exceptional colour fastness. A surface painted with two coats is guaranteed to have a lifetime of between 10 and 15 years.

Nanoseal Wood is an agent for impregnating wood based on nanotechnology, which makes the wood surface strongly water-repellent (www.nanoprotect.co.nz/nanoseal-wood.htm). According to the manufacturer, the water-repellent effect is achieved through structural changes to the surface to a depth of 25 nm. As a result of the treatment, the contact angle for water is as large as 120° so that water runs off the surface instead of penetrating into the wood substrate. The treatment also has a UV-protecting effect.

As a result of the treatment, dirt particles cannot penetrate into the wood and all dirt is therefore washed off by rain or can easily be washed off mechanically. According to the manufacturer of Nanoseal Wood, the product gives a wood surface which is water-repellent, dirt-repellent, mould-resistant, self-cleaning and resistant to weather and UV radiation. According to the manufacturer, the product can be used on painted surfaces and on a new wood substrate provided the substrate is first aged. No information is available concerning its lifetime and how a new surface treatment should be carried out. The product was launched onto the market in 2005 by an Australian company.

NanoWood is, according to the manufacturer a multifunctional water-based silane for the surface treatment of untreated wood surfaces (www.nanoprotect.co.uk/wood-protection.html). Nanowood gives wood and stained wood water- and oil-repellent properties and is very similar to Nanoseal Wood in that a tight cross-linked network is created which, according to the manufacturer, is chemically bonded to the substrate. The product gives a strong and efficient protection against weather and wind, micro-organisms and moisture for up to 4 years.

According to the manufacturer, the advantages of the product are that it is dirt- and water-repellent, that the wood substrate remains permeable to water, that drops of water on the surface dissolve and wash away water-soluble contamination on the surface, that it is transparent so that the texture and colour of the wood are still visible, that it is easy to clean, that it protects against UV and that it reduces the growth of algae, moulds and moss. The product is recommended for the surface treatment of wooden structures such as fences, façades and garden furniture, etc. A product with similar characteristics is marketed by the company Paint Protection Systems (http://www.paintprotectionsystems.com/wood-paints.asp).

The German paint company Caparol has developed a dirt-repellent façade paint *AmphiSilan NQG* with the aid of nanotechnology, where they try to combine the benefits of silicone paints and silicate paints. This gives a paint with both a strong water-repellent effect and good adhesion to the substrate.

AmphiSilan NQG utilizes the so-called NQG technology which employs nanoparticles of quartz which, together with “acrylate tendons” acting as bridges between particles, create a three-dimensional network. The paint gives a strong, shape-stable and nanostructured surface on which it is difficult for foreign particles to fasten. The harder the surface of the paint, the smaller is the amount of dirt which can be retained. If in addition the surface can be rinsed with water which can rapidly dry, at the same time as the penetration of water into the underlying substrate is hindered, the conditions for achieving a clean surface are ideal according to the manufacturer.

The nano-structured surface ensures that rain rapidly forms a very thin film of water which wets the whole surface and gives a thorough cleaning and a rapid drying of the treated wood surface. The appearance is, according to the manufacturer, not spoilt by any patches or streaks from water drops which have dried up or run across the surface. The NQG technique is a further development of the Caparol Clean Concept, which utilizes capillary hydrophobia and photocatalysis and has already been a success for Caparol. Capillary hydrophobia ensures that moisture can easily pass out through the paint layer, while the water-repellent silicone resin layer prevents water from passing in. The combination of the NQG technique and the Caparol Clean Concept gives AmphiSilan NQG a self-cleaning surface. Rainwater does not penetrate through the coating layer, whereas water vapour inside the wood is able to pass out. AmphiSilan NQG also contains photoactive titanium dioxide which breaks down contaminants which cannot be washed away by rainwater. Titanium dioxide has a photocatalytic effect which begins when titanium dioxide reacts photochemically, stimulated by the UV radiation in sunlight. AmphiSilan NQG is recommended for use on well-anchored previously painted substrates such as façades and also on structures of concrete, light-weight concrete, well-bound plaster and cement board. For external wooden façades, the CARAT oil-based paint system utilizes the NQG technique.

Herbol-Symbiotec is a façade paint which is the result of many year's development cooperation between Akzo Nobel and BASF (www.azonano.com/news). The product is based on nanotechnology and the use of the binder Col.9 which has been developed by BASF.

Col.9 is a new type of binder which is a nanocomposite hybrid with organic and inorganic components, Fig. 20, giving a binder combining the properties typical of both inorganic and organic binders. After drying, Col.9 consists of a soft inorganic plastic polymer matrix containing evenly distributed hard inorganic nanoparticles of quartz in a three-dimensional network, where the organic polymer matrix acts as an adhesive.

Col.9 gives a good balance, providing moisture protection together with a sufficiently high moisture vapour permeability. In contrast to the brittle mineral-based coatings, Col.9-based coatings have little tendency to crack. They can also withstand high temperatures and, because of their high silicon content, the coatings have little tendency to develop thermoplastic stickiness. The mineral particles give the coating a hydrophilic surface on which raindrops immediately spread. From a cleaning point of view, this has two advantages: in heavy rain dirt particles are easily washed away from the surface, and the thin moisture film rapidly dries, which inhibits mould growth. Col.9 exhibits a relatively good colour fastness and

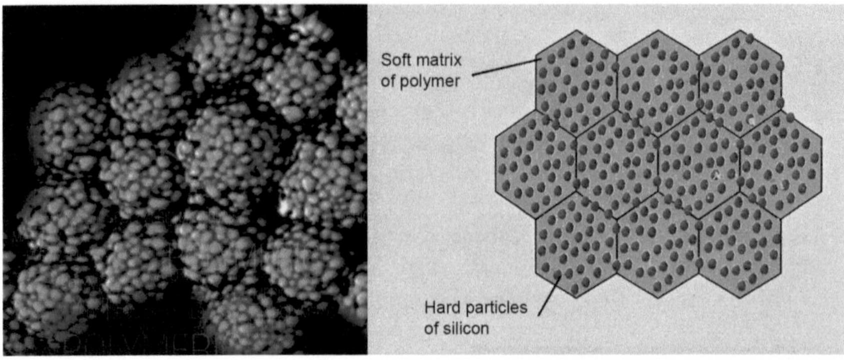

Fig. 20 Col.9 is a nanocomposite binder consisting of 10–20 nm silicon particles in a soft organic polymer matrix

weather resistance. The coating also contains additives which provide protection against mould growth.

The nanotechnology which Col.9 represents is now also being utilized in the development of a paint for the outdoor protection of wood. BASF reports that new binders have been developed with dirt-repellent and sufficient water protection in combination with high water vapour permeability. The first prototype paint for window frames was introduced in 2009.

Recently, various new *Silicon-based products* have entered the market. One example is OrganoWood (www.organoclick.com) that is a surface treatment technique which utilizes silicon molecules to give a wood surface which is more rot-resistant and more resistant to fire. Wood treated with OrganoWood is not classified as hazardous waste and this makes recycling much easier. The product is transparent and does not greatly affect the natural colour of the wood, which means that the timber retains its natural colour but ages and develops a greyish colour with time.

5.8 Environmental Aspects of Coatings

When working with paints and lacquers, a user can be exposed in various ways to potential health risks; inhalation of fumes from evaporating solvents, spray mist or dust from sanding of lacquer that is not fully cured or direct contact with liquid paints and lacquers during handling, either via the skin or eyes.

The most obvious and best-known risk associated with working with paints and lacquers is the threat posed by overexposure to solvents. For this reason, it is important to adopt protective measures in the form of sealed systems and ventilation/air cleaning systems, and to use personal protection, such as various types of masks. Substances other than solvents can also pose a health risk upon inhalation. You should therefore check the health risks stated on the labels when using a product (i.e. warning symbols and risk phrases). The initial effect of inhaling large quantities of organic solvents is a form of inebriation that causes impaired judgement, retarded reactions, clumsiness and tiredness. Extended exposure to concentrated solvent emissions can result in chronic damage to the liver, kidneys and brain. However, not all solvents are equally dangerous. Some of the solvents that pose the greatest danger are aromatic solvents, such as toluene and xylene. To ensure safety at work, public authorities cooperate with medical experts on a continual basis, to determine Occupational Exposure Standards for substances that can be dangerous if inhaled. The legal requirements vary from country to country. Most countries apply limit values (occupational exposure limits) to the substances. These limit values are often expressed as averages over specified time periods, e.g. 8 h or 15 min. Sometimes the limit values must not be exceeded, sometimes they must be observed only if they are technically and economically reasonable. Air pollutants consist of a combination of different substances. Exposure readings taken at the place of work are based on all the substances present in the air, after which the additive hygienic effect can be assessed. The additive hygienic effect (HE) can be summarized: $HE = C1/L1 + C2/L2 + C3/L3$, where C1, C2, C3 are the observed concentrations of substances 1, 2, 3, etc., and L1, L2, L3 are the occupational exposures limit values (OEL values) of these same substances (expressed in the same units). In most countries the HE value must be less than 1 to be acceptable.

In recent years, the risks posed by the inhalation of various types of dust have attracted increasing attention. Such risks are associated primarily with dust that is respirable, i.e. dust particles that are so small that they can penetrate deep into the lungs. The afflictions that can arise, such as asthma or cancer, can be caused by a number of different factors. In some cases, the condition is the result of a toxic reaction, when particles are partially or completely dissolved in the lungs. In other cases, damage to the lungs may derive from a strictly physical injury (e.g. silicosis). Sometimes, health may be damaged when other substances attach themselves to the surface of inhaled particles, and therefore remain trapped in the lungs longer than if they had simply been inhaled in a gaseous form. It is therefore important to protect oneself against overintense exposure to dust, of whatever type. The lists of occupational exposure limits state general hygienic limits for dust and define specific limits for a number of special substances.

A brief description of the labelling system for health and environmental warnings on paints and lacquers is given in Table 5. The most common warning symbols together with risk phrases (R-phrases) are described. Although other symbols and R-phrases exist, these are seldom used for paints and lacquers in the wood finishing industry.

Environmental analyses of wood coatings are complicated. Since currently acceptable LCA methodology leaves ample room for estimates and assumptions that may vary from one practitioner to another, subtle variations in techniques and data sources influence the coating manufacturers' calculations of the life cycles of their products, making it essential to use appropriate techniques to compare their results with each other. Different companies also seem to favour different ways of generating impact scores, such as by weighting VOC emission data by

Table 5 Labelling system for health and environmental warnings on paints and lacquers

Toxicity level	R-phrases	Description
Very toxic/toxic	R26	Very toxic by inhalation
	R27	Very toxic in contact with skin
	R28	Very toxic if swallowed
	R23	Toxic by inhalation
	R24	Toxic in contact with skin
	R25	Toxic if swallowed
	R39	Danger of very serious irreversible effects
	R48	Danger of serious damage to health by prolonged exposure
Carcinogenic	R45	May cause cancer
	R49	May cause cancer by inhalation
Mutagenic	R46	May cause hereditary genetic damage
Toxic for reproduction	R60	May impair fertility
	R61	May cause harm to the unborn child
Harmful	R20	Harmful by inhalation
	R21	Harmful in contact with skin
	R22	Harmful if swallowed
	R42	May cause sensitization by inhalation
	R48	Danger of serious damage to health through prolonged exposure
	R65	May cause lung damage if inhaled
Carcinogenic	R40	Possible risk of irreversible effects
Mutagenic	R68	Possible risk of irreversible effects
Harmful for reproduction	R62	Possible risk of impaired fertility
	R63	Possible risk of harm to the unborn child

photochemical reactivity rather than reporting raw emissions data, further complicating comparisons. One of the challenges encountered is in the data on emission impacts associated with the extraction and refining of natural resources to create some of the raw materials (pigments, additives, etc.). Some help can be achieved from a project carried out in US to develop a platform for accurately accounting for the life cycle impacts of paint and coatings from raw material extraction through to end-of-life (Anon 2011).

6 Aspects on Additives in Waste Wood for Combustion

Wood and wood-based products have unique end-of-life properties. In addition to the recycling of by-products like sawdust, chips and offcuts into particleboard, many other panel products are manufactured from recycled wood. However, beyond this, wood is increasingly being used as a substitute for fossil fuels, providing a renewable energy source which simply returns to the atmosphere the CO_2 it originally removed. Cascading enhances the efficiency of resource utilization by a sequential re-utilization of the same unit of a resource for multiple high-grade material applications followed by a final use for energy conversion (Sirkin and ten Houten 1994). Primary raw materials are thus saved and positive effects due to the substitution of finite materials by renewable resources can be increased (Gustavsson and Sathre 2011). For the recycling of wood and wood fibres it is important that the components, i.e. chemicals or other materials, added to wood in the different process stages do not prevent the use of the wood in the next cycle or prevent the return of the ash to the natural life cycle.

Recovered wood fuel is wood material which has previously had another use and which is to be burned, usually in an energy conversion plant (SIS 2009). It can consist of e.g. demolition wood, shuttering wood, packaging wood or spillage from renovation and new constructions, and it is one component of the wider category of biofuels, which includes liquors from the pulp industry, peat, recycled waste paper, straw, etc.

Recovered wood fuel is generally designated recycled waste wood or, when the material is in a finely divided (chipped) state, chipped recycled waste wood. Recycled wood need not be used solely as a source of energy in a heat-producing plant. It can also be reused or recovered through other processes, e.g. for the manufacture of charcoal. In countries where it is forbidden to use organic waste as landfill (e.g. in Sweden since 2005), there are in principle only two ways of recovering waste wood material (Sundqvist et al. 2009). One is to re-use the material in some type of structure and the other is to use it to produce energy through combustion. According to Tullin and Jermer (1998), approximately half of all recycled waste wood can be reused for building purposes, while the remainder should be used for energy conversion. Most of the installations which utilize recycled chipped waste for energy conversion use a mixed fuel with other solid fuels. The fuel mix usually contains 10–40 % of chipped waste, but there are also plants which use 100 % chipped waste wood (Andersson and Tullin 1999).

6.1 Recycling of Wood

The increasing scarceness of primary raw materials leads to a greater focus on secondary resources. Deposits from urban infrastructure, mainly the building stock, are potentially a major secondary resource. Höglmeier et al. (2013) analysed the amounts of wood incorporated in the building stock of south-east Germany, and calculated resulting streams of recovered wood in order to quantify potentially available volumes for an environmentally beneficial cascading utilization of these secondary resources. They found that considerable amounts of recovered wood in a condition suitable for a resource-efficient use in cascades can be expected to originate from the building stock: 26 % of the recovered wood is suitable for reuse and 27 % could be channelled into other high-value secondary applications.

In several European countries, recycled waste wood is not only free from energy and carbon dioxide taxes but also qualifies for an electricity certificate, which means that recycled waste wood is relatively in demand as a fuel. On the other hand, it makes greater demands on the plant which burns recycled waste wood than on a plant which uses only wood fuel directly from silvicultural raw material (so-called virgin wood fuel), and this means that many choose not to use recycled waste wood in their fuel mix (Strömberg 2005). In the long term, when the demand for recyclable ash for the nutritional compensation of e.g. forest land increases, a fuel mix containing recycled waste wood may be less and less attractive, since the ash from the combustion of recycled waste wood usually contains substances which must not be spread in nature.

A major problem in the recovery of energy from recycled waste wood is that some of the material has been treated in some way, e.g. impregnated with a wood preservative or subjected to a surface treatment. Nor is it unusual for other structural components of, e.g. plastic or metal to accompany the recycled waste wood (Tullin and Jermer 1998; Strömberg 2005). The contamination in recycled waste wood can be divided into chemical contaminants derived from, e.g. paint and impregnating materials, and mechanical contaminants such as plastic, fittings, glass, nails and screws (Jermer et al. 2001; Strömberg 2005). Mechanical contaminants are in general easier to deal with than chemical contaminants, since they can be separated from the recycled waste wood before combustion either at the demolition site or in subsequent handling (waste sorting), but this usually leads to increased costs.

The contaminants in the recycled waste wood vary strongly depending on the type of wood recovered and also on how it is handled in connection with demolition, sorting and chipping. The chemical contaminants may include, e.g. potassium, chlorine, sodium, zinc and lead if paints or other surface treatments have been used. Other possible contaminants are copper, chromium and arsenic if the wood has been impregnated with a wood preservative (Andersson and Tullin 1999). According to e.g. Jermer et al. (2001), most of the chemical contaminants can be found in the fine fraction of the material obtained after chipping, and this means that it is therefore possible to separate much of the contamination from the purer fractions. Since the fine fraction can amount to 25–40 per cent by weight of the chips this may mean, however, that a large fraction of the chipped waste wood must then be dealt with in some other way, e.g. by deposition as landfill.

The chipped recycled waste wood is characterized by a high ash content, up to ca. 25 % has been reported, and a moisture content of 20–35 % (Andersson and Tullin 1999), and the ash from the chipped waste contains large amounts of heavy metals (Tullin and Jermer 1998). In addition, the nitrogen and chlorine contents are in certain cases slightly higher in recycled waste wood than in virgin wood fuel (Berg et al. 2003). Nitrogen influences the combustion process negatively with regard to e.g. discharges. There is a risk that chlorine in combination with certain metals, particularly copper, may lead to the formation of dioxins. Dioxins are a joint name for ca. 200 chlorine-containing substances which may give rise to cancer, foetal damage, infertility and various skin diseases (The National Encyclopaedia 2015).

The combustion of chipped recycled waste wood involves more engineering problems than the combustion of virgin wood fuel. The risk of coating and corrosion in the combustion chamber of the recovery boiler increases with high contents of zinc and chlorine (Berg et al. 2003; Strömberg 2005). In boilers with a fluidized bed, it is primarily the ash feed out and the bottom nozzles which become clogged when the ash fuses together. In grate-fired boilers, the problem is that the air vents in the grate become clogged because the metals in the fine fraction melt at the high temperatures which arise locally on the grate. According to Strömberg (2005), it is better to burn chipped waste wood in a boiler with a fluidized bed than in a grate-fired boiler because the coatings in the digester as well as the flue ash contain much less zinc when a fluidized bed is used. By increasing the sulphur and silicate content in the combustion, the problem can be further reduced.

The use of recycled waste wood as fuel makes high demands on the exhaust gas cleaning in the plant, particularly the equipment for particle separation, to fulfil the increasingly strict emission regulations. With regard to heavy metals in the flue gas, any differences between various fuel assortments are due chiefly to differences in the efficiency of the particle separation equipment at the plant (Andersson and Tullin 1999). The main problem is that heavy metals are enriched in the ash. According to measurements reported by Berg et al. (2003), the emission limits regarding metals from boilers are met if the regulations regarding particle separation are complied with. This is because heavy metals such as arsenic, cadmium, lead and zinc are enriched in the fly ash particles which are separated from the flue gases.

The discharge of hydrochloric acid and sulphur dioxide is somewhat higher in the combustion of chipped recycled waste wood than in the combustion of virgin wood fuel. The discharge of gaseous substances is, however, more difficult to remedy, since this requires flue gas treatment in addition to particle removal. Conventional flue gas condensation is often sufficient to reduce the discharge of hydrochloric acid, but further steps are needed to meet the requirements regarding sulphur dioxide.

To meet the increasingly restrictive emission limits in many European countries in the incineration of CCA-impregnated wood, both flue gas condensation and an efficient particle removal are required (Sundqvist et al. 2009). In the combustion of wood impregnated with modern copper-based wood preservatives, it is, however, sufficient if the combustion unit is equipped with an electrostatic precipitator.

6.1.1 Chemical Contaminants in Recycled Waste Wood Due to Surface Treatment

Surface treatments such as varnish, paint and mould or rot inhibitors are the most common contaminants in recycled waste wood, and since these compounds are enriched in the ashes from the combustion of the recycled waste wood, these ashes should not be returned to the natural eco-cycle but should instead be deposited or taken care of in some other way.

When wood is impregnated with a wood preservative or is surface treated in some way, both organic and inorganic substances are introduced into the wood, but it is primarily the inorganic substances which lead to problems in the combustion process since they are retained in the ashes, the fly ashes and the flue gases, while the organic substances are degraded by the combustion (Tullin and Jermer 1998). If flue gas cleansing is incorporated into the combustion plant, most of the inorganic contaminants will finally end up in the ashes and this will reduce the possibility of returning the ashes to the natural eco-cycle.

Paints consist primarily of binder, solvent, pigment and various additives such as siccatives, fillers and fungicides and some these additives can lower the melting point of the ashes and thus increase the risk of coating and clogging in the boiler.

The binders in the paint normally cause few problems in combustion since they are usually organic compounds consisting primarily of carbon, hydrogen and oxygen, but there are exceptions. Protein-based binders contain nitrogen and they thus increase the content of "fuel nitrogen" which leads to increased problems with the emission of nitrogen oxides (NO_x). Some varnishes also contain both nitrogen and sulphur and this leads to increased amounts of NO_x and SO_2 in the flue gases.

The solvents in modern paints are in general not a problem, since they are either water-based or organic. Older types of paint can, however, contain chlorine-based solvents and these can lead to an increased content of chlorine in the combustion of chipped recycled waste wood, but the content should not be particularly high since the solvents in paints normally evaporate when the paint dries.

Pigments can be divided into two categories, viz. organic and inorganic. Of these, it is the inorganic pigments which are of greatest concern since they contain oxides of metals such as zinc, titanium, chromium, iron and in some cases also lead. Certain inorganic pigments also contain nitrogen. The main causes of problems in organic pigments are nitrogen and chlorine, but some so-called organic pigments also contain metals such as copper and chromium. The presence of nitrogen leads to the emission of NO_x, and chlorine in the fuel contributes to the formation of hydrochloric acid and dioxins during combustion.

Of the various paint additives, the siccatives are the most interesting from a contaminant viewpoint since they may contain heavy metals, the most common being zinc, calcium, lead, iron, manganese, lithium, barium and cobalt. Fillers can contain calcium, silicon, aluminium, magnesium and barium. Together with alkali metals, silicon can form compounds which melt at a low temperature. Water-based paints must also be pH-adjusted since they otherwise tend to be too acidic. Sodium hydroxide is used for this purpose, and this introduces sodium which can be a problem in the combustion plant.

6.1.2 Chemical Contaminants in Recycled Waste Wood as a Result of Impregnation

Previously, the substances predominantly used for the impregnation of wood were CCA, which contaminates the wood with primarily chromium, copper and arsenic, and creosote. Until the middle of the 1990s, CCA was the dominant substance for the impregnation of sawn wood, but its use in EU has since been forbidden for use above ground and CCA has been replaced by water-based copper-containing wood preservatives (Sundqvist et al. 2009). The contaminants in these agents are mainly copper, nitrogen and chlorine. Between 1975 and 1995, most of the wooden window frames made in Sweden were impregnated with a preservative containing organic tin compounds, Nowadays, wooden window frames are impregnated with triazoles which are organic compounds containing nitrogen and chlorine.

Creosote is a distillate obtained from coal tar and it is organic in nature which means that creosote-impregnated timber is well suited for combustion (Tullin and Jermer 1998; Jermer et al. 2001). Apart from the fact that creosote is derived from a fossil material, it is similar to normal wood fuel from an environmental point of view. Creosote is used mainly for the impregnation of railway sleepers and telegraph poles. Creosote predominates as a wood preservative for sleepers, and it is only very rarely that sleepers are impregnated with any other preservative agent (Jermer et al. 2001). In the future, a long-term environmental adaptation of wood preservative substances will be necessary so that wood impregnated with preservatives can be recycled with the smallest possible effect on the environment.

6.1.3 Chemical Contamination Due to Other Additives

In addition to substances for surface treatment and impregnation, a number of other additives used in the manufacture of wood products will be present in recycled waste wood and may give rise to problems in the combustion process. These are primarily the glues used in various structural elements (glued-laminated timber) and in board materials such as MDF, plywood and particleboards. Nitrogen, sulphur and chlorine are important components in many adhesives. As previously stated, nitrogen leads to the emission of NO_x, sulphur to the emission of SO_2 and chlorine in combination with certain catalytic metals, particularly copper, gives rise to dioxins.

6.2 The Need for Recycled Waste Wood in Heat-Generating Plants

Chipped recycled waste wood is an excellent fuel for energy recovery as long as the plant has sufficient flue gas cleansing and provided the ash is handled correctly. The ash can in some cases be used to cover landfill if the limiting value for heavy metals is not exceeded.

The quality demands for recycled waste wood that is to be incinerated are similar within European countries, but they vary from country to country. All combustion plants generally wish to have waste wood sorted at source without plastic, metal or surface treatment and with a small fine fraction. In certain countries (e.g. Sweden) a special licence is required for the combustion of CCA-impregnated timber, but the restrictions are milder in the case of modern preservatives and such recycled timber can be incinerated in most plants. Increasing energy requirements, stricter EU-directives for the classification of fuel waste, and increasing deposition taxes are expected to lead to a greater demand for recycled waste wood sorted at source and to a greater need for methods for the quality assurance of waste wood. If recycled waste wood is to be incinerated in normal combustion plants and not in special installations, a sorting at source of the raw material at the point of demolition is necessary, and this is not normally the case in, e.g. Sweden. In Germany, sorting at source of the waste wood is very common and impregnated and painted waste wood is incinerated in special plants while the clean waste wood material is incinerated in conventional biofuel boilers. It is not therefore uncommon in Germany for chipped waste wood to have a content of contaminants at the same low level as virgin wood fuel (Berg et al. 2003).

The ash from incinerated recycled waste wood that contains surface-treated, impregnated or otherwise contaminated wood must, as already mentioned, not be returned to nature because of its content of heavy metals. In order to be able to return the ash from the combustion of recycled waste wood, the waste wood must be sorted at source so that the clean chipped waste wood can be incinerated in a conventional plant, while the impregnated and surface-treated wood is incinerated in a special plant with sufficient flue gas cleansing, after which the ash can be deposited as landfill.

Sorting at source can take place either at the demolition site or through a fractionation of the recycled waste wood so that the fine fraction containing most of the contamination can be separated. The fine fraction is then incinerated in a plant which has a licence to do so. Since large quantities of the contaminants which lead to problems in the boiler are present in the fine fraction, the methods for fractionation of the waste wood could be adapted so that a large part of this fraction can be separated from the chipped waste wood.

7 Concluding Remarks

Today, forestry and forestry-related industries are sharply in the focus of discussions concerning the major challenges for the future. A great challenge for humankind is to develop a sustainable society. Such a society requires the use of renewable materials, a large reduction in the use of non-renewable natural resources and a large reduction in environmental impacts, including a drastic reduction of greenhouse gas emissions. This includes methods and resources used in industry, construction and consumer products, including those made from wood.

One way of reducing the emission of carbon dioxide is to use a larger proportion of wood products and to increase the lifetime of these products so that the carbon is stored over a longer period of time. Another possibility is to replace energy-intensive materials with wood and wood-based products. Forest resources and wood products can, therefore, play an important role in a long-term strategy for sustainable development and lessen man's impact on the environment. Replacing non-renewable materials with wood-based ones is hence crucial for the development of a sustainable society, and new knowledge is required to show how resource-efficient processes for wood products with a low environmental impact can be designed and implemented.

Although sawn timber and wood-based products remain vital from an economic and utilitarian standpoint, it is increasingly essential that the technology to create EWPs from wood and other renewable resources be understood, applied and further developed. In the future, wood products also have to be refined so that all the life cycle phases—production, operation, retrofitting and end-of-life—are considered and optimized as a whole, including the energy and material chains from forest to final services.

Wood is to a great extent combined with additives to achieve a more efficient use of the wood resource. Additives from renewable resources have a great potential in the future, due not only to environmental concerns but also to the anticipated future scarcity of petroleum and petroleum-based products, but economic and durability questions have to be solved. Bio-based adhesives already have a commercial and industrial impact in, e.g. plywood and particleboard production, and the adhesive is never more than 10 % by weight of the whole composite panel. This is sufficient to conform to the performance and costs required by the wood panel industry and their respective product standards. Additives from renewable resources may in the future have the greatest potential over other additives because of their easiness in reusing, recycling and recovery for energy conversion.

References

Andersson C, Tullin C (1999) Förbränning av returflis – kvalitetssäkring och drifteserfarenheter (The combustion of recycled wood—quality assurance and operational experiences), Report No 668. Thermal Engineering Research Association (Värmeforsk), Stockholm

Anon (2011) Life cycle assessment of volatile organic compounds (LCA-VOC) in paints & coatings: final report. National Center for Manufacturing Sciences Inc, Ann Arbor

Barthlott W, Ehler N (1977) Raster-Elektronenmikroskopie der Epidermis-Oberflächen von Spermatophyten (Scanning electron microscopy of the epidermal surfaces of spermatophytes). Tropische und subtropische Pflanzenwelt 19:110 (Akad Wiss Lit Mainz). Franz Steiner Verlag, Mainz

Barthlott W, Neinhuis C (1997) Purity of the sacred lotus, or escape from contamination in biological surfaces. Planta 202(1):1–8

Berg M, Andersson A, Andersson C et al (2003) Förbränning av returträflis – Etapp 2 av ramprojekt returträflis (Combustion of recovered wood—stage 2 of the framework project recovered wood), Report No 820. Thermal Engineering Research Association (Värmeforsk), Stockholm

Berger J (2006) Bauplatte oder der gleichen, deren herstellung und werwendung (Building panel or the like, and production and use thereof). PCT Patent PCT/AT2006/000141, 6 Apr 2006

Bobacz D (2002) In CNC-Technik gefertigte zimmermannsmäßige Verbindungsmittel: Untersuchung des Schwalbenschwanzzapfens. Diplomarbeit, Universität für Bodenkultur, Vienna

Boonstra MJ, Pizzi A, Zomers F et al (2006) The effects of a two stage heat treatment process on the properties of particleboard. Holz Roh Werkst 64(2):157–164

Bourgois J, Guyonnet R (1988) Characterisation and analysis of torrefied wood. Wood Sci Technol 22:143–155

Carrasco F, Roy C (1992) Kinetic study of dilute-acid prehydrolysis of xylan-containing biomass. Wood Sci Technol 26:189–208

Carus M, Gahle C (2008) Injection moulding with natural fibres. Reinf Plast 52(4):18–25

CEN (1992) EN-120, Wood-based panels. Determination of formaldehyde content: extraction method (called the perforator method). European Community for Standardization (CEN), Brussels

CEN (2007) CEN/TS 15679, Thermal modified timber—definitions and characteristics. European Community for Standardization (CEN), Brussels

CEN (2013) EN-301, Adhesives, phenolic and aminoplastic, for load-bearing timber structures—classification and performance requirements. European Community for Standardization (CEN), Brussels

Clemons CM (2002) Wood–plastic composites in the United States—the interfacing of two industries. Forest Prod J 52(6):10–18

Dagher HJ, Bragdon MM, Lindyberg RF (2002) Advanced fiber-reinforced polymer-wood composites in transportation applications. Transp Res Rec 1814:237–242

Detlefsen WD (2002) Phenolic resins: some chemistry, technology and history. In: Chaudhury MK, Pocius AV (eds) Adhesive science and engineering: surfaces, chemistry and applications. Elsevier, Amsterdam (chap. 20)

DIBt (2012) Allgemeine bauaufsichtliche Zulassung DIBT Z-9.1-649 für Schwalbenschwanz-Verbindung in Bauteilen (General building inspection approval DIBt Z-9.1-649 for dovetail joint in components). Deutsche Institut für Bautechnik, Berlin

Dietrichs HH, Sinner H, Puls J (1978) Potential of steaming hardwoods and straw for feed and food production. Holzforschung 32:193–199

Dietsch P (2005) Development of a finite-element model for parameter studies of a dovetail connection. Diplomarbeit, Fachgebiet Holzbau, Technische Universität, München

DIN (1979) DIN Standard 7707: types of resin impregnated and compressed laminated wood and insulating wood. Deutsches Institut für Normung

Dunky M (2003) Adhesives in the wood industry. In: Pizzi A, Mittal KL (eds) Handbook of adhesive technology. Marcel Dekker, New York (chap. 47)

Ek M, Gellerstedt G, Henriksson G (eds) (2000) Pulp and paper chemistry and technology, vol 1–5. Walter de Gruyter GmbH & Co, Berlin

Ellis S, Paszner L (1994) Activated self-bonding of wood and agricultural residues. Holzforschung 48:82–90

Feifel S, Poganietz WR, Schebek L (2013) The utilization of light weight boards for reducing air emissions by the German wood industry—a perspective? Environ Sci Eur 25(5). doi:10.1186/2190-4715-25-5

Frihart CR (2011) Wood adhesives. Vital for producing most wood products. For Prod J 61(1):4–12

Frihart CR, Birkeland MJ (2014) Soy properties and soy wood adhesives. In: Brentin RP (ed) Soy-based chemicals and materials. ACS Sym Ser 1178, American Chemical Society, Washington, DC, pp 167–192

Frybort S, Mauritz R, Teschinger A, Müller U (2008) Cement bonded composites—a mechanical review. BioResources 3(29):602–626

Fuch W (1928) Genuine lignin. I. Acetylation of pine wood. Ber Dtsch Chem Ges 61B:948–951

Geimer RL, Leao A, Armbruster D, Pablo A (1994) Property enhancement of wood composites using gas injection. In: Maloney TM (ed) Proceedings of the 28th Washington State University International particleboard/composite materials symposium, April 12–14. Pullman, pp 243–259

Goldstein IS (1955) The impregnation of wood to impart resistance to alkali and acid. For Prod J 5:265–267

González-García S, Feijoo G, Heathcote C, Kandelbauer A, Moreira M (2011) Environmental assessment of green hardboard production coupled with a laccase activating system. J Cleaner Prod 19(5):445–453

Graubner W (1992) Encyclopedia of wood joints. Taunton Press, Newtown

Gustavsson L, Madlener R, Hoen HF et al (2006) The role of timber material for greenhouse gas mitigation. Mitig Adapt Strat Glob Change 11(5/6):1097–1127

Gustavsson L, Sathre R (2011) Energy and CO_2 analysis of wood substitution in construction. Clim Change 105(1/2):129–153

Haller P (2007) Concepts for textile reinforcements for timber structures. Mater Struct 40:107–118

Haller P, Wehsener J, Werner TE, Hartig J (2013a) Recent advancements for the application of moulded wooden tubes as structural elements. In: Aicher S, Reinhard HW, Garrecht H (eds) Material and joints in timber structures. Springer, Heidelberg, pp 99–108

Haller P, Putzger R, Wehsener J et al (2013b) Formholzrohre – Stand der Forschung und Anwendungen (Molded wood pipes—state of research and applications). Bautechnik 90:34–41

Hallström S (1995) Glass fibre reinforcement around holes in laminated timber beams. Department of lightweight Structures, Report No 95-14. Royal Institute of Technology, KTH, Stockholm

Hein PRG, Aparecida de Sá V, Bufalino L et al (2009) Calibrations based on near infrared spectroscopic data to estimate wood-cement panel properties. BioResources 4(4):1620–1634

Hillis WE (1975) The role of wood characteristics in high temperature drying. J Inst Wood Sci 7:60–67

Hochstrate M (2000) Untersuchungen zum Tragverhalten von CNC gefertigten Schwalbenschwanzverbindungen (Investigations on the structural behavior of CNC crafted dovetail joints). FH Hildesheim/Holzminden/Göttingen, Hildesheim

Holzner H (1999) Entwicklung eines Nachweisverfahrens zur Bemessung von speziellen (maschinell gefertigten) Zapfenverbindungen (Development of a detection method for the design of special (machine-made) mortise and tenon joints). Institut für Tragwerksbau-Fachgebiet Holzbau, Technische Universität Münich, Germany

Hood EE, Nelson P, Powell R (eds) (2011) Plant biomass conversion. Wiley-Blackwell, West Sussex

Horn O (1928) Acetylation of beech wood. B Dtsch Chem Ges 61B:2542–2545

Höglmeier K, Weber-Blaschke G, Richter K (2013) Potentials for cascading of recovered wood from building deconstruction: a case study for south-east Germany. Resour Con Recy 78:81–91

IEC (2006) IEC Standard No 61061-1, Non-impregnated densified laminated wood for electrical purposes—part 1: definitions, designation and general requirements. International Electrotechnical Commission, Geneva

Insulander R (1997) The Fenno-Urgian two-wood bow—a missing link. Bull Primitive Technol 14(2):35–39

Jermer J, Ekvall A, Tullin C (2001) Inventering av föroreningar i returträ (Inventory of pollutants in recycled wood). Report No 732, Thermal Engineering Research Association (Värmeforsk), Stockholm

Kamdem DP, Pizzi A, Triboulot MC (2000) Heat-treated timber: potentially toxic by-products presence and extent of wood cell degradation. Holz Roh Werkst 58(4):253–257

Kamke FA (2013) THM—a technology platform or novelty product? In: Medved S, Kutnar A (eds) Characterization of modified wood in relation to wood bonding and coating performance. Proceedings of the COST FP0904 and FP1003 International Workshop, Rogla, 16–18 Oct 2013

Killen G (2000) Wood—procurement and primary processing. In: Nicholson PT, Shaw I (eds) Ancient Egyptian materials and technology. Cambridge University Press, Cambridge, pp 353–368

Kitek Kuzman M, Kutnar A (2014) Contemporary Slovenian timber architecture for sustainability. Green energy and technology. Springer, Berlin, p 163

Klauditz W, Stegmann G (1955) Beiträge zur Kenntnis des Ablaufes und der Wirkung thermischer Reaktionen bei der Bildung von Holzwerkstoffen (Contributions to the knowledge of the sequence and the effect of thermal reactions in the formation of wood materials). Holz Roh Werkst 13:434–440

Klyosov AA (2007) Wood-plastic composites. Wiley-Interscience, Hoboken

Knight EV, Wulpi M (eds) (1927) Veneers and plywood. Their craftmanship and artistry, modern production methods and present-day utility. The Ronald Press Company, New York

Koehler A, Pillow MY (1925) Effect of high temperatures on the mode of fracture of a softwood. Southern Lumberman 121:219–221

Kollmann FFP, Kuenzi EW, Stamm AJ (1975) Principles of wood science and technology II. Wood based materials. Springer, New York

Kreuzinger H, Spengler R (1999) Zum Tragverhalten von maschinell abgebundenen Zapfenverbindungen aus Konstruktionsvollholz zwischen Haupt-und Nebenträger (For structural behavior of machined hardened tenon joints from construction timber between the main and secondary beams). Untersuchungsbericht LKI, 7313, Technische Universität Münich, Germany

Kuroki Y, Nagatomi W, Yamada J (1993) Manufacture of light-weight cement-bonded particleboard. In: Moslemi AA (ed) 3rd International inorganic-bonded wood and fiber composite materials conference, Spokane 28–30 Sept 1992

Kutnar A, Burnard MD (2014) The past, present, and future of EU wood adhesive research and market. In: International conference on wood adhesives. Toronto, 9–11 Oct 2014

Kutnar A, Hill C (2014) Assessment of carbon footprinting in the wood industry. In: Muthu SS (ed) Assessment of carbon footprint in different industrial sectors, vol 2., EcoProductionSpringer, Berlin, pp 135–172

Kutnar A, Sandberg D, Haller P (2015) Compressed and moulded wood from processing to products. Holzforschung 69(7):885–897

Lande S, Westin M, Schneider M (2004) Properties of furfurylated wood. Scand J For Res 19 (5):22–30

Lauer JP (1933) Fouilles du service des antiquités à saqqarah (secteur nord, Novembre 1932-mai 1933). (Excavations of the Department of Antiquities at Saqqara (north sector, November 1932–May 1933)). Annales du service des antiquités de L'Égypte 33:155–166

Li K, Peshkova S, Gen X (2004) Investigation of soy protein-kymene® adhesive systems for wood composites. J Am Oil Chem Soc 81:487–491

Linné Cv (1737) Flora Lapponica (The flora of Lapland). Schouten, Amsterdam

Lopez-Anido R, Xu H (2002) Structural characterization of hybrid FRP-glulam panels for bridge decks. J Compos Constr ASCE 6(3):194–203

Lopez-Anido R, Michael AP, Sandford TC (2003) Experimental characterization of FRP composite-wood pile structural response by bending tests. Mar Struct 16(4):257–274

Lopez-Anido R, Muszynski L, Gardner DJ, Goodell B, Herzog B (2005) Performance-based material evaluation of fiber-reinforced polymer-wood interfaces in reinforced glulam members. J Test Eval 33(6):385–394

Lucas A (1936) The wood of the third dynasty, ply-wood coffin from Saqqara. Annales du service des antiquités de L'Égypte 36:1–4

Mahapatra K, Gustavsson L (2008) Multi-storey timber buildings: breaking industry path dependency. Build Res Inf 36(6):638–648

Maloney TM (1977) Modern particleboard and dry-process fibreboard manufacturing. Miller-Freeman Publications, San Francisco

Navi P, Sandberg D (2012) Thermo-hydro-mechanical processing of wood. EPFL Press, Lausanne

Neinhuis C, Barthlott W (1997) Characterization and distribution of water-repellent, self-cleaning plant surfaces. Ann Bot 79(6):667–677

Näsström J (2005) Stjärnsågning igen – Nystart i mindre skala och med ny teknik (Star-Sawing again—a restart on a smaller scale and with new technology). Nordisk träteknik Såg & Trä 9:12–13

Phleps H (1982) The craft of log building. Lee Valley Tools Limited, Ottawa

Pizzi A (2006) Recent developments in eco-efficient bio-based adhesives for wood bonding: opportunities and issues. J Adhes Sci Techn 20(8):829–846

Pizzi A, Mittal KL (2003) Handbook of adhesive technology. Marcel Dekker, New York, p 672

Pokharel N (2003) Behaviour and design of sandwich panels subject to load buckling and flexural wrinkling effects. PhD thesis, School of Civil Engineering, Queensland University of Technology

Rebstock F, Bomark P, Sandberg D (2015) Makerjoint, a new concept for joining members in timber engineering—strength test and failure analyses. Pro Ligno 11(4):397–404

Richter K (2001) LCA—reuse/recycle. In: Johansson CJ, Pizzi T, van Leemput M (eds) Wood adhesion and glued products, Report on the state of the art of COST Action E13, pp 161–180

Robeller C, Mayencourt P, Weinand Y (2014a) Snap-fit joints-CNC fabricated, integrated mechanical attachment for structural wood panels. In: ACADIA 2014 design agency: Proceedings of the 34th annual conference of the association for computer aided design in architecture. Los Angeles, 23–25 Oct 2014, Riverside Architectural Press, pp 189–198

Robeller C, Hahn B, Mayencourt P, Weinand Y (2014b) CNC-gefräste Schwalbenschanzzinken für die Verbindung von vorgefertigten Bauteilen aus Brettsperrholz (CNC machined dovetail joints for the connection of prefabricated components made of laminated timber). Bauingenieur 89:487–490

Rowell RM (2005) Handbook of wood chemistry and wood composites. Taylor & Francis, Boca Raton

Rowell RM, Simonson R, Tillman AM (1986) A simplified procedure for acetylation of chips for dimensionally stabilized particleboard products. Pap Puu 68(10):740–744

Sandberg D, Haller P, Navi P (2013) Thermo-hydro and thermo-hydro-mechanical wood processing: an opportunity for future environmentally friendly wood products. Wood Mat Sci Eng 8(1):64–88

Savastano Júnior H, Warden PG, Coutts RSP (2000) Brazilian waste fibers as reinforcement of cement-based composites. Cement Concr Compos 22(25):379–384

Schefferus J (1673) Lapponia. ex officina Christiani Wolffii, Frankfurt am Main, Germany (See also translation by John Scheffer (1674). The history of Lapland, Oxford)

Schindler C (2009) Ein architektonisches Periodisierungsmodell anhand fertigungstechnischer Kriterien, dargestellt am Beispiel des Holzbaus (An architectural periodization based production engineering criteria, using the example of timber construction). Eidgenössische Technische Hochschule Zürich, ETH, Zürich

Shade N (1998) The strip-built sea kayak. Ragged Mountain Press, Camden

Sirkin T, ten Houten M (1994) The cascade chain: a theory and tool for achieving resource sustainability with application for product design. Resour Con Recy 10:213–277

SIS (2009) SS 187106: Fasta biobränslen och torvbränslen – Terminologi (Solid boifules and peat fuels—Terminology). Swedish Standards Institute, Stockholm

Skuratov N (2010) New lightweight solid wood panels for green building. In: Proceedings of the International convention of society of wood science and technology and United Nations Economic Commission for Europe—Timber Committee, Geneva, 11–14 Oct 2010

Sonti SS, GangaRao HVS (1996) Banding timber crossties using composite fabrics for improving their performance. Materials for the new technology. ASCE Press, Washington, DC, pp 1449–1457

Stamm AJ, Seborg RM (1939) Resin-treated wood. Ind Eng Chem 31:897–992

Stamm AJ, Burr HK, Kline AA (1946) Staybwood. Heat-stabilized wood. Ind Eng Chem 38:630–634

Stevens WC, Turner N (1970) Wood bending handbook. HMSO, London

Strömberg B (2005) Bränslehandboken (Fuel Handbook). Report No 911, Thermal Engineering Research Association (Värmeforsk), Stockholm

Suchsland O, Woodson GE (1986) Fiberboard manufacturing practices in the United States. USDA Forest Service Agriculture Handbook, 640, US Government Printing Office, Washington, DC

Suida H, Titsh H (1928) Chemistry of beech wood: acetylation of beech wood and cleavage of the acetyl-beech wood. Ber Dtsch Chem Ges 61B:1599–1604

Sundqvist JO, Erlandsson M, Solyom P, Högberg B, Bergman G (2009) Impregnerat trä i kretsloppet – rekommendationer för restprodukthantering (Chemical preservative wood in recycling—recommendations for waste management). Report No B1827, Swedish Environmental Research Institute IVL, Stockholm

Tarkow H, Stamm AJ, Erickson ECO (1946) Acetylated wood. Report No 1593, USDA, Forest Service, Forest Prod Lab, Madison

Tannert T, Prion H, Lam F (2007) Structural performance of rounded dovetail connections under different loading conditions. Can J Civ Eng 34(12):1600–1605

Tannert T, Lam F, Vallée T (2011) Structural performance of rounded dovetail connections: experimental and numerical investigations. Eur J Wood Wood Prod 69(3):471–482

Tiemann HD (1915) The effect of different methods of drying on the strength of wood. Lumber World Rev 28:19–20

The National Encyclopedia (2015) http://www.ne.se/dioxiner. Accessed 24 Aug 2015

Tullin C, Jermer J (1998) Inventering och energiutvinning av träskyddsbehandlat virke i Sverige (Inventory and energy of preservative-treated wood in Sweden). Report No 653, Thermal Engineering Research Association (Värmeforsk), Stockholm

Wallenberg FT, Bingham P (2010) Fiberglass and glass technology, energy-friendly compositions and applications. Springer Science+Business Media LLC, New York

Werner F, Richter K (2007) Wood building products in comparative LCA. A literature review. Int J Life Cycle Ass 12(7):470–479

Wilson TRC (1920) The effect of kiln drying on the strength of airplane woods. Report No 68, National Advisory Committee for Aeronautics, Washington, DC

Zwerger K (2012) Wood and wood joints: building traditions of Europe and Japan. Walter de Gruyter, Berlin

Forest-Based Bioproducts Used for Construction and Its Impact on the Environmental Performance of a Building in the Whole Life Cycle

Roman Kunič

Abstract Because of the increasing emissions of various pollutants, rapidly growing energy demands, and possible global warming consequences, renewable and nature friendly construction materials are gaining importance. Climate change due to global warming, variations in oil prices and environmental threats have led to significant demand for the wood bio and bio-based products, where construction products represent a significant volume. Construction industry has significant environmental, social and economic impacts on the society. The wood buildings require much lower process energy and result in lower carbon emissions than the buildings of other materials such as brick, aluminium, steel and concrete. If a shift is made towards greater use of wood in buildings, the low fossil fuel requirement for manufacturing wood compared with other materials is much more significant in the long term than the carbon stored in the wood building products. To compare the environmental performance of a building in the whole life cycle, the carbon footprint of 15 insulation materials was calculated. The obtained values were compared to the actual effect of the respective thermal insulation. The transparency of the comparison was achieved by taking into account the specific weight of each material, as well as differences in their thermal conductivity (λ). Moreover, a study of environmental neutrality of different types of insulation materials is presented. For the thermal insulation materials with the lowest environmental impact, the environmental neutrality is reached in only 0.57 heating season and in 7.89 heating seasons for the insulation with the highest environmental impact.

Keywords Buildings · Carbon dioxide · Construction · Energy · Greenhouse gases · Timber · Wood · Wood products

R. Kunič (✉)
Faculty of Civil and Geodetic Engineering, University of Ljubljana,
Jamova cesta 2, 1000 Ljubljana, Slovenia
e-mail: roman.kunic@fgg.uni-lj.si

A. Kutnar and S.S. Muthu (eds.), *Environmental Impacts of Traditional and Innovative Forest-based Bioproducts*, Environmental Footprints and Eco-design of Products and Processes, DOI 10.1007/978-981-10-0655-5_5

1 Introduction

The impact that the society has on environment is becoming more intense. Resource consumption is getting beyond the absorption capacity of the environment. Since the beginning of the industrial revolution the world economy has relied on the exploitation of relatively cheap energy, raw materials and other resources. In addition to the lack of cheap energy and other resources and the uncertainty of supply of key fossil fuels (oil and natural gas), the world is faced with the problem of climate change, with huge amounts of waste and increasing pollution. Sustainable development is necessary to stabilize resource supplies, and should become a responsibility of each company, the public sector and the general public alike. Only with sustainable development it will be achieved that the prices of products and services will also cover the costs of environmental impacts throughout their life cycle. To achieve this, the practices of energy saving, environmental planning, repair, recycling, reuse and production of materials and finished products are being developed and implemented by world's economies (Kitek Kuzman and Kutnar 2014). Because of increasing emissions of various pollutants, rapidly growing energy demands, and possible global warming consequences, renewable and nature friendly construction materials are gaining importance. Additionally, buildings in their lifetime globally consume 40 % of material and energy, and produce as much waste (Kunič 2007). Therefore, a greater emphasis has to be given to the conservation of non-renewable energy sources and reduction of the consumption of raw materials and other resources, in concert with the recycling and waste management at the end of the life cycle of a building (Košir et al. 2010).

Forest-based bioproducts, mainly wood, as a source material for construction are similar as solar energy in energy sector, the cleanest renewable natural resource available, and does not produce air or other polluters. Forest wood is produced with the help of permanent solar radiation that arrives at the Earth's surface, which is not constant at certain location, but rather the opposite: it does vary significantly according to the time of day, time of year and weather conditions. During the photosynthesis process, wood grows by using CO_2 from the air. Due to this process, CO_2 is stored in wood until the final release through mechanisms such as burning oxidation, rotting, bacterial or fungal decays or consumption by insects finish its life cycle. Wood and forest-based bioproducts in built houses and their constructions are ideal for storing CO_2 for long periods—decades or even centuries. Further on, after its life cycle wood can be reused or eventually used for energy purposes.

It is certain that wood will remain one of the leading construction materials in the future due to its aesthetic, structural, and environmental characteristics. Today, the use of wood in architecture is becoming fashionable. It is believed that developments in wood products and timber construction will shape the future of sustainable development. Wood is still one of the most accessible materials, we can smell, hear, touch, and see (it is pleasing to the eye) natural wood. A similar development can be seen in the facades of modern structures, where imitation wood is increasingly being used.

In construction technology, we generally differentiate between wood, timber and lumber. Definitions used in this chapter are (Dehne and Krueger 2006):

- Wood: the hard, fibrous, lignified substance lying under the bark of trees. It consists largely of cellulose and lignin. Wood is a natural material and is irregular by nature.
- Timber: the wood of trees cut and prepared for use as building material (e.g. beams and posts).
- Lumber: timber cut into marketable boards, planks or other structural members of standard or specified length.

Timber is a wholly sustainable construction material for the developed world, as it re-grows, can be sustainably harvested, is fully recyclable, and stores carbon. Using certified wood ensures that the timber is supplied from sustainable forestry, and not from developing countries, where precious 'old-growth' forests are illegally logged (Lehmann 2013).

In this chapter, wood as raw material for construction elements and thermal insulation is being presented. Carbon footprint of thermal insulation materials in building envelopes is given as comparison among the fossil-based and wood-based thermal insulation materials. The reuse, upgrading and recycling of construction elements after life cycle of a building are discussed including the environmental impacts of buildings in the whole life cycle.

2 Wood as Raw Material for Construction Elements

Timber has excellent construction features. Its compressive strength is almost equal to that of concrete, but its tensile strength is significantly higher. The most important advantage over concrete is its much lower weight. Moreover, if the weight of both materials is equivalent, timber satisfies almost the same construction requirements as steel. Nevertheless, on account of its relatively low value of the modulus of elasticity, which is approximately three times lower than that of concrete and twenty times lower than that of steel, timber is not suitable for structures with extreme spans or heights, although it has become an ever more frequently used material in multi-storey prefabricated construction (Premrov and Žegarac-Leskovar 2013).

Wood is a commonly used construction material in many parts of the world because of its reasonable cost, ease of working, attractive appearance, and adequate life if protected from moisture and insects. The various species of wood have a number of physical characteristics that enable performance of wood needed in building construction. However, wood properties vary among species, between trees of the same species, and between pieces from the same tree. This leads to variability in the performance of wood, which is one of its inherent deficiencies as a material (Dinwoodie 2000). However, wood composites that reduce the variability of wood as a material can be produced.

The purpose of wood composites is the transformation of materials into items of greater value. The ideal wood-based composite is either stronger, lighter, more durable, attractive, or cheaper than competing materials. The basic element for composite wood products can be timber, veneers, veneer strands, flakes, strands, wood particles, fibre, adhesives and other additives, and finishing materials.

Wood-based composites have long been used as both decorative and structural components in the human environment. These materials extract the best properties of wood (and eliminate or minimize the defects) and combine them with other materials (adhesives, plastics, etc.) to create a wide variety of new products that meet market demands. A broad range of wood-based composites have been developed in the past. Wood composites are a family of materials, which contain wood either in whole or fibre form as the basic constituent (Bodig and Jayne 1982). A binding adhesive of either natural or synthetic origin interconnects the wood or fibre elements. Composites are normally thought of as two-phase systems, i.e. particles interconnected by a binder; wood composites, however, are multiphase systems including moisture, voids and additives. Furthermore, Berglund and Rowell (Berglund and Rowel 2005) defined a composite as two or more elements held together by a matrix. By this definition, what we call "solid wood" is also a composite. Solid wood is a three-dimensional composite composed of cellulose and hemicelluloses (with smaller amounts of inorganics and extractives), held together by a lignin matrix (Kitek Kuzman and Kutnar 2014).

In Europe, the most commonly produced wood-based panels are particleboard and Medium-Density Fibreboard (MDF). However, Oriented Strand Board (OSB), traditional plywood, insulation board, and hardboard are also important products. Other more recent products include Laminated Veneer Lumber (LVL), light MDF (LDF), High-Density Fibreboard (HDF) and Cross-Laminated Timber (CLT). In the past years, technological innovations have advanced the field of wood-based panels. Most notably, hot pressing and the consequent viability of thermosetting resins have improved composites produced from particles and strands (particleboard, OSB), fibres (as MDF, HDF) and veneers (plywood, LVL). Sawn softwood timber is most commonly used directly in structural applications or as a component of engineered products (e.g. Glue-Lams). Planed (also called surfaced or dressed) timber has been machined to have a smooth, uniform surface and ensures proper sizing. Air-dried timber has been dried without energy aid, while kiln-dried timber has been dried with energy aid, often using cogenerated electricity or natural gas as an energy source to provide heat and maintain regular air flow.

The performance of composites can be tailored to the end-use application of a product by optimally arranging the physical configuration of the wood components, adjusting the density, varying the resin type and amount and incorporating additives to increase water or fire resistance or to resist specific environmental conditions (Kitek Kuzman and Kutnar 2014).

2.1 Wood Structural Composites

Timber has recently been used in a much more sophisticated way than previously. In recent years, approaches to timber construction which constructions are characterized by the method of assembling prefabricated structural elements and the structure of the facade (envelope). General building systems consist of similar wall and ceiling elements, though these elements can also be used in combination to form a building structure; for example, the application of solid wood elements in ceilings of framework structures (Lehmann 2013). Most common structural composites are glued laminated timber (Glulam), OSB, plywood and cross-laminated timber (CLT).

Glued laminated timbers (Gluelam) are structural composite beams used to support large loads in building construction. Sawn timber, selected for stress-related mechanical properties, are glued and arranged in layers (with the high-grade timber in the outer layers, and low-grade timber in the inner layers) with the grain direction parallel to the length of the timber. The size of the resulting glued laminated timbers may vary greatly, allowing the beams to be used as needed for a specific application. Glued laminated timbers for indoor use may use adhesives that are less resistant to the effects of the outdoor environment (e.g. relative humidity and temperature), while glued laminated timbers for outdoor use must use adhesives that are more resistant to changes in the outdoor environment (Kitek Kuzman and Kutnar 2014).

OSB is a structural panel product most often used for roof, wall, and floor sheathing in construction. The product is usually made of three or more layers with strands in each layer oriented in alternating directions (i.e. parallel to the length of the panel, or perpendicular to it). Water-resistant adhesives are used for OSB. The strands in the outer layer are oriented with the grain direction parallel to the length of the panel. The strands used are typically about three times longer than they are wide.

Plywood is made from thin layers of wood, which has been peeled from a log on a rotary lathe. These thin veneers are then combined into three or more (usually an odd number) of layers in alternating grain directions. The outer layers are aligned with the grain direction parallel to the length of the panel. Plywood for indoor applications may use an adhesive that is less water resistant than plywood for outdoor use. In indoor applications plywood is often used in furniture. Plywood for outdoor applications must use a water-resistant adhesive. Sheathing is the most common use of plywood in exterior applications.

Beginning around 2000, CLT, a new type of wood-based panel, began development, and soon after went into commercial production. CLT is an engineered wood panel typically consisting of three, five, or seven layers of dimension lumber oriented at right angles to one another and then glued to form structural panels with exceptional strength, dimensional stability, and rigidity. Adhesives based on polyurethane basis are used. Modular cross-laminated timber (CLT, also called cross-lam) panels form the basis of low carbon, engineered construction systems using solid wood panels that can be used to build residential infill developments of 10 storeys or higher. Multi-apartment buildings of 4–10 storeys constructed entirely with timber can be built. The potential of prefabricated engineered solid wood panel systems, such as CLT, as a sustainable building material and system is only just being realized around the globe. CLT panels are a 'value-added' timber product that can substitute for concrete or steel, which are both very carbon intensive. As layers of timber boards are glued crosswise (width of wood stripes usually varies between 80 and 240 mm, with thickness between 10 and 40 mm), the load-bearing capacity of CLT panels in different directions is increased for taking up compression loads, while the shrinkage and swelling as a result of humidity variations is eliminated. Design for disassembly principles means that buildings can be disassembled or adapted more easily as use dictates (Lehmann 2013).

CLT elements can be used for load-bearing shear walls, wall columns, lintel beams and floors. These products are expected to play an important role in the future for single- and multi-storey timber buildings (Kitek Kuzman and Kutnar 2014).

The main advantages of the new engineered timber system, which moves towards low-carbon construction solutions, are that it offers new ways of constructing affordable housing, which will benefit from the many advantages CLT panels offer. Lehmann (2013) lists among the advantages of CLT panels:

- over 50 % reduction of construction time, significantly minimizing construction costs;
- high material efficiency, zero-waste construction
- less weight (lightweight, better handling on-site and smaller foundations. Timber has a weight of 500 kg/m^3, compared to concrete with 2450 kg/m^3);
- earthquake resistant (ideal for all cities on fault lines);
- strong and durable, locally sourced, leading to distinctive architecture;
- thermally efficient, with low embodied energy;
- recyclable and biodegradable, made from an abundant and fully renewable wood resource (sustainably harvested pine plantation forestry);
- negative net carbon emissions, sourced from sustainably grown and harvested forest stock

Since timber is the only material that has the capacity to store carbon in large quantities over a long period of time, solid wood panel construction offers the opportunity to turn buildings into 'carbon sinks'. Thus, the historically negative environmental impact of urban development and construction can be turned around with CLT construction on brownfield sites. Wegner and his colleagues noted that 'the energy budgets of products and buildings made of wood show that they may use less energy over their total life cycle (manufacture, use, maintenance and disposal) than can be recovered from the waste products of their production and from their recycling potential at the end of their life cycle: they are energy-positive. No other construction material is so comprehensively energy-efficient and therefore climate effective as wood' (Wegener et al. 2010).

2.1.1 Environmental Impacts of Wood Products

Only recently there is more research on sustainable forest management, the life cycle of wood products and the use of timber in construction in regard to CO_2 emission cuts. For instance, new research reports that instead of keeping the wood to grow in forests for hundreds of years, it is beneficial if the wood is regularly harvested and used in construction to replace concrete or steel that consumes more fossil fuel during manufacturing; in this way, a multi-fold level of carbon dioxide can be removed from the atmosphere (Lippke et al. 2004). The analysis has identified a number of possibilities for the use of wood to replace concrete and steel products that utilise fossil fuel resulting in one-way generation of carbon dioxide. According to Lippke et al. (2004), sustainably managed forests practically offer two-way flow of carbon dioxide: they absorb the carbon dioxide during their growth and when the tree dies and decays, falls back or burns; carbon goes back to the atmosphere, thus remaining carbon neutral. Therefore, the researchers recommend a quicker growth of trees and cropping of the wood before the trees become less active, to utilise them in place of steel or concrete in construction (Lehmann 2013).

The products with the lowest carbon footprints are air-dried sawn timber and glued laminated timber. This is unsurprising, because these products are processed less than wood-based composites. The glued laminated timber has higher carbon footprint due to adhesives, but is still negative. Wood has a negative footprint because of the carbon dioxide fixed by the original living tree. The emissions associated with harvesting, transporting, and processing sawn wood products are small compared to the total amount of carbon stored in the wood. This means that even when energy use for harvesting, transport and processing are taken into account, sawn wood still has a negative footprint. Wood-based composite production requires additional energy inputs to process raw materials, manufacturing by-products, and recycled wood into the desired form, as well as adhesives and other additives to form the composite matrices, which considerably increases the carbon footprint of these wood products. The highest carbon footprint among the compared products has plywood for outdoor use, followed by MDF and particleboard (Kitek Kuzman and Kutnar 2014).

In order to quantify the effect of carbon storage in wood products, it is necessary to estimate the life of the products. The carbon emissions may occur within a few days of harvest, or after many years of use in wood products or as paper or cardboard decaying very slowly in landfills. Some studies allocate products to three or four categories (e.g. very short, short and long-life products) (Karjalainen et al. 1994, 1999; Schlamadinger and Marland 1996). Other studies use an annual decay rate (Maclaren and Wakelin 1991; Winjum et al. 1998). The average life of fuel wood and waste has been taken as one year, paper products three years and solid wood products 40 years represent a good approximation (Buchanan and Bry Levineb 1999).

There has been some attempt to deal with the evaluation of biogenic carbon storage in long-life products in national standards. In UK, this issue was dealt with in Publically Available Specification (PAS) 2050 (2011). Additionally, methodologies for evaluating the atmospheric carbon stored in products are given in the International Reference Life Cycle Data (ILCD) Handbook, published by the

European Commission Joint Research Centre (Institute for Environment and Sustainability). In both, a 100-year assessment period is considered, following IPCC guidelines. It is recommended that fossil and biogenic carbon releases (as CO_2 and CH_4) should be differentiated. Two methods for calculating the weighted average of the effect of carbon storage in a product are given, although for a product with a life less than 2 years, no carbon storage benefit can be assigned. This can only be applied to the storage of biogenic carbon, which is assigned a negative CO_2 value. However, this cannot be applied if the biogenic carbon is derived from old growth or native forests, where land-use change has occurred (Kitek Kuzman and Kutnar 2014).

2.2 *Wood as Raw Material for Thermal Insulation*

Since buildings are the key player in the use of land resources, especially in the time of their operation, it is essential to focus on improving their energy efficiency. The high dependence on imported energy sources gives further urgency to this effort. The basic action needed to increase the energy efficiency of buildings is to use a proper insulation material in the building envelope. Due to increasing environmental awareness, the thickness of the thermal insulation in the building envelope has increased on the average, in new constructions as well as in case of retrofitting. Investment costs of thick thermal insulation layers are higher, but the operating costs are significantly lower, even to the extent that highly insulated building is cost-effective throughout its life cycle (Kunič and Krainer 2009). As it does not require any additional changes in the construction project, using the thicker insulation is by far the most important and most (cost) effective investment leading to energy savings (Kunič and Krainer 2010; Vattenfall 2015).

Insulation in the building envelope can be made of a variety of insulating materials, which differ in their chemical composition (organic and inorganic), their origin (derived from new raw materials, partially or totally recycled), their specific weights (minimum 12 kg/m^3 up to approximately 600 kg/m^3), their thermal conductivity (λ ranges from less than 3 mW/(m K) to over 45 mW/(m K)), and in the resistance to physical (moisture, elevated temperature, presence of UV radiation, pressure, shear, laminated and other strength) and chemical factors (the presence of organic solvents, moisture, oxidation, burning, etc.).

3 Carbon Footprint of Thermal Insulation Materials in Building Envelopes

This section delivers carbon footprint comparison among 15 most often used thermal insulation materials: EPS, 'grey' reflective EPS (with infra-red reflector additives), XPS, PU—polyurethane, low-density glass wool, high-density glass wool, rock wool of two different densities, wood wool of two different densities, recycled cellulose, cork, foamed glass, aerogel, and VIP. The materials differ in density (16–380 kg/m^3, relation 1:24) and thermal conductivity (6 mW/(m K) to 90 mW/(m K), relation 1:15), as is shown in Table 1 and graphically presented in Figs. 1 and 2. Detailed description (chemical and physical composition) of the listed insulation materials is available in technical documentation of the materials' manufacturers and various literature sources (Pfundstein et al. 2008).

Review of reports given in promotional documents of insulation materials showed that the biggest difference is in citing the environmental impact per unit weight of the product "per se", which means that this comparison does not take into account the differences in specific weights or different values of thermal conductivity (λ).

Following the common Life Cycle Assessment (LCA) methodology (ISO 14040:2006; PAS 2011), the scope and goal of the study was to compare the environmental impact of selected thermal insulation materials. Environmental impact was analysed with the "Cradle-to-gate" variant, an assessment of a partial product life cycle from manufacturing ('cradle') to the factory gate (i.e. before it is transported to the consumer). The use phase and disposal phase of the product were omitted. The calculations included emissions caused by the installation, but did not include the possible emissions caused during the service life of a building, its operational phase. Since the analysed insulation materials might differ in service life, the results of the analysis that would account for service life of insulation materials might be different. However, reliable data of service life of analysed insulation materials were not available and were therefore not included in the calculations.

The environmental burdens associated with each insulation material were considered from raw material acquisition, through the manufacturing/processing stages, accounting for the production and use of fuels, electricity and heat, as well as taking into account transportation/distribution impacts at all points along the product supply chain. Functional unit for the calculation was determined to be 1 kg of the specific thermal insulating material. Based on the determined goal and scope of the study, the life cycle inventory of input/output data for the LCA calculations was compiled. Data of energy inputs, raw materials, products, co-products, waste, and releases to air, water and soil and the upstream life cycle impacts of input materials were not analysed specifically for this project. Instead, sound secondary life cycle data were sourced from Ecoinvent database 3.0. (Ecoinvent 2013). Table 2 shows a list of data that were used in carbon footprint calculations. The data collected were modelled in Simapro software environment: SimaPro Analyst Indefinite, Ecoinvent v2, Product Ecology Consultants, PEC (Ecoinvent 2013; Simapro 2009). Emissions and consumptions were translated into environmental effects, which were grouped and weighed. Carbon footprint was calculated with methodology IPCC 2001 GWP 100a V1.02 (Climate Change 2001), which contains the climate change factors of IPCC with a timeframe of 100 years. Emissions, appearing in the future, 100 years after the start of the process, were taken into account. IPCC characterization factors for the direct (except CH_4) global warming potential of air emissions were used.

Table 1 Density, thermal conductivity and required thickness of different thermal insulation materials to obtain thermal transmittance $U = 0.20$ W/(m^2 K); the data are compiled from years of experience in construction practice

	Density (ρ)			Thermal conductivity (λ)			Required thickness of thermal insulation for thermal transmittance (U-value) of 0.20 W/(m^2 K) for most commonly used densities and thermal conductivities of thermal insulations for building purpose (cm)
	From (kg/m^3)	To (kg/m^3)	Most commonly used for building envelopes (kg/m^3)	From (mW/(m K))	To (mW/(m K))	Most commonly used for building envelopes (mW/(m K))	
EPS	10	30	16	34	45	37	17.5
EPS with reflective additives	12	28	16	30	35	32	15.2
XPS	28	45	32	31	44	38	18.0
PU polyurethane	28	100	45	22	30	25	11.8
Glass wool—low density	15	40	22	34	40	36	17.0
Glass wool—high density	40	150	80	30	45	38	18.0
Rock wool—low density	20	120	70	33	42	40	18.9
Rock wool—high density	120	200	155	35	48	45	21.3
Wood fibre wool—low density	50	270	120	38	60	50	23.7
Wood fibre wool—high density	350	600	380	75	110	90	42.6
Cellulose—recycled	30	80	60	40	50	44	20.8
Cork	100	220	160	45	70	50	23.7
Foam glass	90	200	170	38	80	60	28.4
Aerogel	60	160	140	13	24	17	8.0
VIP	150	300	170	3	11	6	2.8

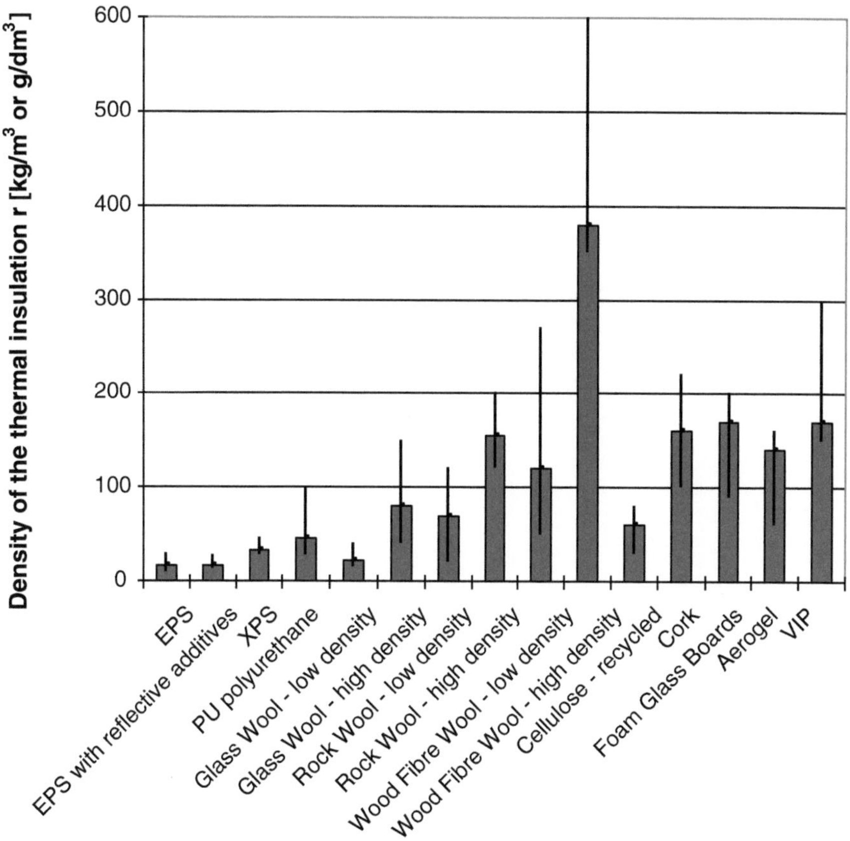

Fig. 1 Density ρ (kg/m^3 or g/dm^3) of the different thermal insulations. *Blue bars* represent average most commonly used thermal insulations for building purpose, whereas *black lines* (error bars) represent variations of densities or thermal conductivities of different types of insulation materials used for other purposes (i.e. industry or other technical use)

They do not include indirect formation of dinitrogen monoxide from nitrogen emissions, do not account for radiative forcing due to emissions of NOx, water, sulphate, etc., in the lower stratosphere as well as upper troposphere, do not consider the range of indirect effects given by IPCC (Climate Change 2001), and do not include indirect effects of CO emissions.

Carbon footprints of different insulating materials were calculated by taking into account the thickness of the thermal insulation that is required for fulfilling the condition of the heat transfer (*U*-value) 0.20 W/(m^2 K). Additionally, carbon footprint of thermal insulation materials was compared with that of other building materials. The comparison was performed between the environmental impacts caused by the insulation of the entire exterior building envelope (the average equivalent area of 400 m^2) with the impact on the environment of other building

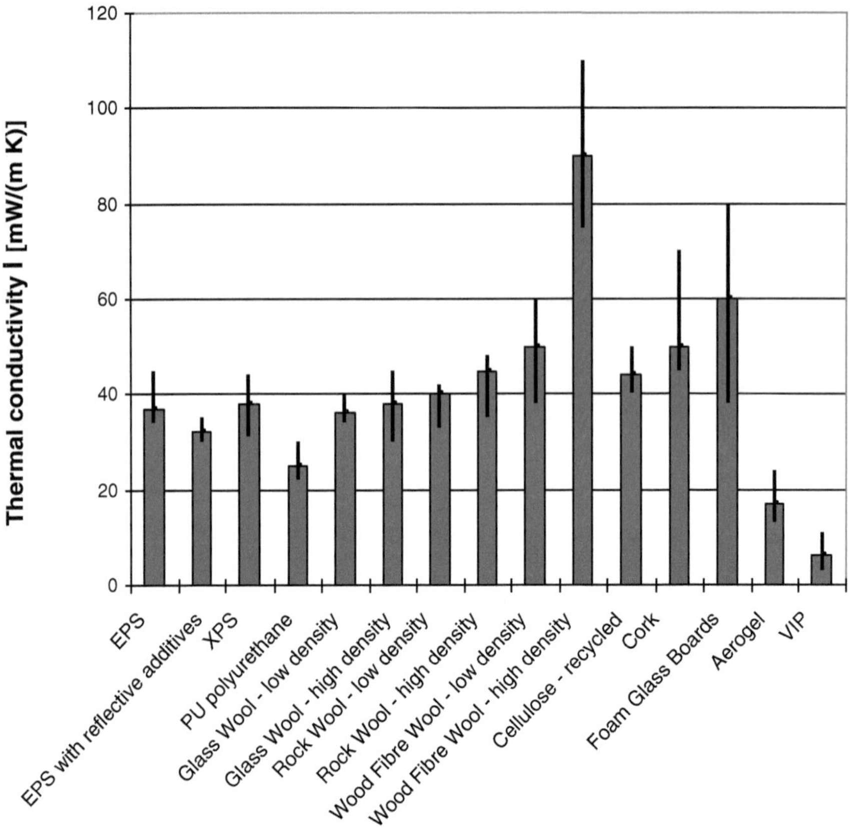

Fig. 2 Thermal conductivity λ (mW/(m K)) of the different thermal insulations. *Red bars* represent average most commonly used thermal insulations for building purpose, whereas *black lines* (error bars) represent variations of densities or thermal conductivities of different types of insulation materials used for other purposes (i.e. industry or other technical use)

materials (reinforced concrete, fired clay bricks, flat window glass, PVC, stainless steel and aluminium elements) (Zabalza et al. 2009).

The study was focused on the evaluation of the environmental impacts of insulation in the external envelope of the building. With the aim to perform transparent comparison, only the environmental impact of the insulation was considered and the environmental impact of all the other elements, such as bearing wall, protective construction, fasteners, adhesives, finishes and plasters, was neglected. As a justification, it has to be noted that the majority of the selected systems of façades, walls and roofs use very similar methods of attachment and composition, irrespective of the type of heat-insulating material. The results of the evaluation are therefore applicable to any construction complex in the external building envelope (outer wall, flat or pitched roofs, floors on the ground, etc.).

Table 2 List of qualitative data for the calculation of the carbon footprint

Input data	
Mass balance of the materials used	Weight of the finished product Weight of each of the materials used in the product Source of individual materials Weight of waste material
Energy	Electricity consumption per product Consumption of fuel (diesel, heating oil, gasoline, wood, etc.) per product
Water	Water consumption per product (drinking water and process water) The amount of waste water in the product and handling Emissions of pollutants in water (type and quantity) per product
Other waste and emissions	Emissions of pollutants into the air (type and quantity of pollutants) per product Other waste materials associated with the production processes, e.g. filters, ash, etc.
Transport	Transport type of transport vehicles for the transport of materials Distances The proportion of unused (empty) vehicles and the proportion of unused return travel
Packaging	Packaging weight and type of material for the protection and packaging

Although heat losses of the building occur through outer walls, ceilings, floors, thermal bridges, other conduction thermal transfer, radiation and ventilation (natural or forced), only conductive thermal losses through external building envelope during heating (winter) season were taken into account. Construction supporting the model external envelopes was assumed to be made of reinforced concrete wall with a thickness of 15 cm, which was on one side plastered with 2 cm thick cement-lime plaster.

Calculations were made for Ljubljana (capital city of Slovenia) with degree days method (3300 K day) (PURES 2010).

Specific annual energy requirement for heat losses through external building envelope, neglecting heat efficiency of inner space heating system, was calculated by

$$E_{\mathrm{BE\,a}} = \mathrm{DD}.U_{\mathrm{BE}} \tag{1}$$

where

$E_{\mathrm{BE}\,a}$ specific annual energy requirements for heat losses through external building envelope (J/(m^2 year))

DD degree days (K. day = 86,400 K s)

U_{BE} thermal transmittance of external building envelope (W/(m^2 K))

The total thermal transmittance of the external building envelope was expressed as

$$U_{BE} = [R_i + R_{BEtot} + R_{INS} + R_e]^{-1} \quad (2)$$

where

R_i thermal resistance of inner air (convection) (m^2 W)/K
R_{BEtot} total thermal resistance of external building envelope (m^2 W)/K
R_{INS} thermal resistance of thermal insulation (m^2 W)/K
R_e thermal resistance of external air (convection) (m^2 W)/K

Thermal resistance of thermal insulation layer

$$R_{INS} = d_{INS}/\lambda_{INS} \quad (3)$$

where

d_{INS} thermal insulation thickness (m)
λ_{INS} thermal conductivity of insulation (W/(m K))

Heat losses through external wall were determined as:

$$q_{BE} = U_{BE}.\Delta T \quad (4)$$

where

q_{BE} heat losses through unit area of external building envelope (W/m^2)
ΔT thermal difference (K)

Carbon footprint of the heat losses through the unit area of external building envelope was taken into account as an average value of available energy sources as 0.250 kg CO_2-eq./(kW h) (or 0.0694 kg CO_2-eq./MJ) (STAT 2012).

3.1 Carbon Footprint

In Fig. 3, carbon footprints of various insulating materials according to their weight, without taking into account different densities and differences in the thermal conductivity of the material (λ), are shown. As for the specific effect of thermal insulation, with the same levels of thermal transmittance (U) to be achieved, different amounts of a particular material are needed. Therefore, the comparison in Fig. 3 cannot serve as the criterion to determine thermal insulating materials with low environmental impact. The analysis should consider the difference in density of each thermal insulation material, as well as differences in their thermal conductivity.

Densities of thermal insulation materials for construction use vary from 16 to 380 kg/m^3 (Table 1). Also the values of thermal conductivity of different insulating

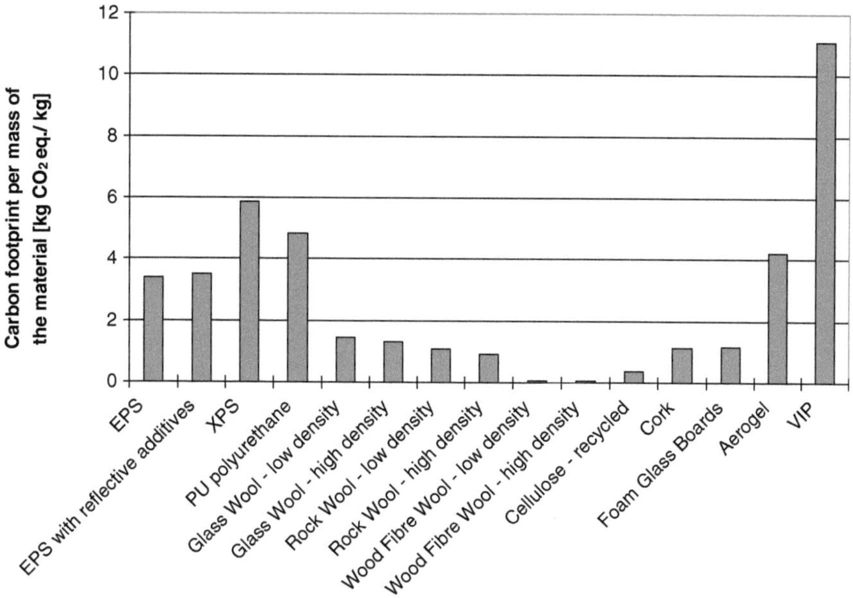

Fig. 3 Carbon footprint of different thermal insulation materials per kilogram mass of the selected material

materials are different (Table 1). The differences in the values of thermal conductivity are mostly in the range of 32 to 90 mW/(m K), with the exception of insulation materials aerogel and vacuum insulation panels (VIP), which have a significantly lower value of thermal conductivity.

Actual environmental impact, expressed in terms of CO_2-eq. emissions for various thermal insulation materials included in the analysis, should therefore be compared based on the same level of thermal transmittance (U) value to be achieved. The necessary thickness of individual type of thermal insulation material, as well as their mass per unit area needed to achieve the thermal transmittance of the outer building envelope of $U = 0.20$ W/(m^2 K) differ. Therefore, the carbon footprint of the selected thermal insulation materials per unit area (m^2) that is needed to achieve thermal transmittance $U = 0.20$ W/(m^2 K) is given (Table 3). These values allow valid comparison and should be used when decisions over thermal insulation are taken based on the smallest environmental impact.

In Fig. 4, the carbon footprint per unit area of the envelope (m^2) is shown, while in Fig. 3, the carbon footprint with respect to the unit mass of the insulating material is presented. The maximal environmental impact caused by the installation of thermal insulation in the building envelope in order to achieve thermal transmittance value of 0.20 W/(m^2 K), is 56.0 kg CO_2-eq. per square metre of the building envelope surface.

Comparison of carbon footprints of the thermal insulation materials (Fig. 3) show that wood-based insulations, in this case wood wool, cause minimal

Table 3 Comparison between the physical properties of thermal insulation materials and carbon footprint of various thermal insulations enabling thermal transmittance $U = 0.20$ W/(m^2 K) per unit area of the building envelope (m^2)

Thermal insulation material	Carbon footprint per mass of most commonly used material for building envelopes (kg CO_2-eq./kg)	Required weight of thermal insulation per surface (1 m^2) for $U = 0.20$ W/(m^2 K) (kg/m^2)	Carbon footprint of thermal insulation per surface unit (1 m^2), for $U = 0.20$ W/(m^2 K) (kg CO_2-eq./m^2)
EPS	3.38	2.80	9.5
EPS with reflective additives	3.50	2.42	8.5
XPS	5.86	5.76	33.7
PU polyurethane	4.83	5.33	25.7
Glass wool—low density	1.46	3.75	5.5
Glass wool—high density	1.30	14.39	18.7
Rock wool—low density	1.08	13.26	14.3
Rock wool—high density	0.90	33.02	29.7
Wood fibre wool—low density	0.06	28.41	**1.7**
Wood fibre wool—high density	0.06	161.92	9.9
Cellulose—recycled	0.37	12.50	4.6
Cork	1.15	37.88	43.6
Foam glass	1.16	48.29	**56.0**
Aerogel	4.20	11.27	47.3
VIP	11.08	4.83	53.5

environmental impact. Recycled cellulose, most often newsprint, also has low impact on the environment. In both products, wood wool and recycled cellulose, a significant part of the carbon footprint is due to the additives that prevent rot, decay and burning. Mineral, glass and stone wool (especially of low density) also have low impact on the environment.

The embodied emissions analysed do not include any offset for the carbon stored in the wood. Approximately 50 % of dry timber is elemental carbon; thus, 1 kg of wood contains approximately 0.50 kg of carbon, which equates to 1.83 kg of CO_2-eq. (Buchanan 2006). Whether to include the carbon stored in cellulose-based materials (and to a far lesser extent small amounts of carbon stored in other materials) when calculating carbon footprint is a much debated issue (Sinha and Kutnar 2012).

In view of the overall environmental acceptability and the use of resources and environment, synthetic or plastic materials have a high environmental impact, compared to natural materials (e.g. wood wool).

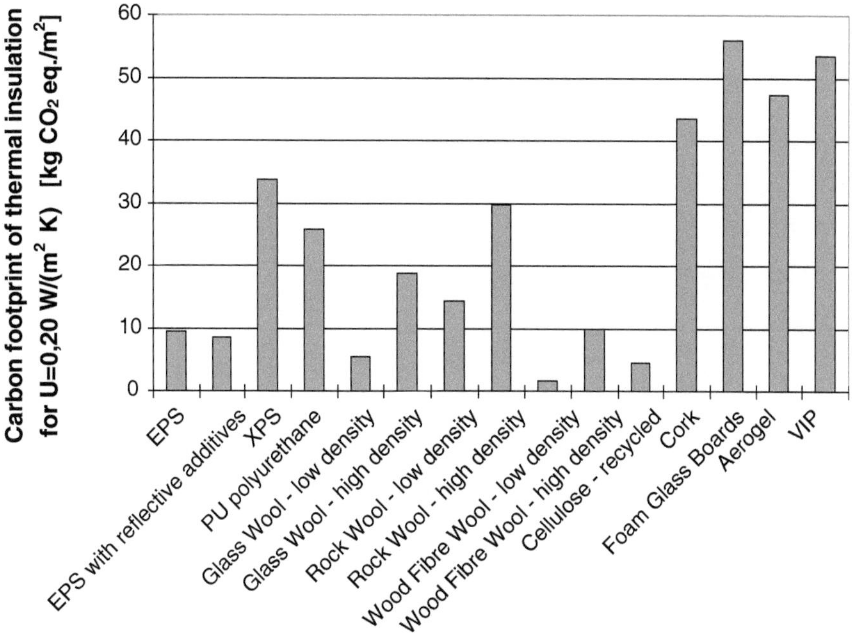

Fig. 4 Carbon footprint of thermal insulation materials needed to achieve the value of thermal insulation of the building envelope U = 0.20 W/(m^2 K); presented per unit area of building envelope (m^2)

The analysis determined that thermal insulating materials with the greatest environmental impact are extruded polystyrene, polyurethane foam, foam glass and mineral wool of high density (Fig. 4). The reason is mainly their relatively high density and consequently large mass of material required to achieve a certain degree of thermal insulation, as well as in many cases their manufacturing processes that have high impact on the environment.

The environmental impact caused by the insulation of the entire external building envelope, assumed to be 400 m^2, with thermal transmittance U = 0.20 W/(m^2 K), compared to the impact on the environment of other building materials, is small. That is true even in the case of thermal insulations with high carbon footprint, such as polyurethane foam, which causes the environmental impact of 10.29 t CO_2-eq. This footprint equates to 57.5 tonnes of reinforced concrete (Table 4). However, it has to be emphasized that an individual building of 400 m^2 contains about three times as much concrete. Furthermore, the same impact is made by 38.0 metric tonnes of fired clay bricks (an average, individual house contains about 4 times as much), or 2412 kg of PVC products, or 6743 kg of steel products or only 1201 kg of aluminium products.

In Table 5, carbon footprint due to the installation of various thermal insulations in the building envelope per unit area (m^2) to achieve the requirements of U-value = 0.20 W/(m^2 K) is presented and compared to the environmental balance

Table 4 Carbon footprint of different thermal insulation materials needed to achieve heat transfer of $U = 0.20$ W/(m^2 K) in 400 m^2 of the whole external building envelope with the carbon footprint of other construction materials

Thermal insulation material	Carbon footprint[a] (t CO_2-eq.)	Equivalent carbon footprint as with reinforced concrete (t)	Equivalent carbon footprint as with mortar (t)	Equivalent carbon footprint as with construction brick (t)	Equivalent carbon footprint as with flat construction window glass (t)	Equivalent carbon footprint as with polyvinyl-chloride (PVC) (t)	Equivalent carbon footprint as with construction steel (t)	Equivalent carbon footprint as with construction aluminium (t)
EPS	3.79	21.17	15.72	13.98	5.45	0.89	2.48	0.44
EPS with reflective additives	3.39	18.96	14.08	12.52	4.88	0.80	2.22	0.40
XPS	13.49	75.39	55.99	49.80	19.42	3.16	8.84	1.57
PU polyurethane	10.29	57.49	42.70	37.97	14.81	2.41	6.74	1.20
Glass wool—low density	2.19	12.23	9.09	8.08	3.15	0.51	1.44	0.26
Glass wool—high density	7.48	41.81	31.05	27.62	10.77	1.75	4.90	0.87
Rock wool—low density	5.73	31.99	23.76	21.13	8.24	1.34	3.75	0.67
Rock wool—high density	11.89	66.41	49.33	43.87	17.11	2.79	7.79	1.39
Wood fibre wool—low density	0.69	3.87	2.88	2.56	1.00	0.16	0.45	0.08
Wood fibre wool—high density	3.95	22.07	16.39	14.58	5.68	0.93	2.59	0.46
Cellulose—recycled	1.82	10.19	7.57	6.73	2.63	0.43	1.20	0.21
Cork	17.42	97.33	72.29	64.29	25.07	4.08	11.42	2.03
Foam glass	22.41	125.18	92.98	82.68	32.24	5.25	14.68	2.61
Aerogel	18.93	105.76	78.55	69.85	27.24	4.44	12.41	2.21
VIP	21.41	119.60	88.83	79.00	30.80	5.02	14.03	2.50

[a]Carbon footprint for 400 m^2 of external whole building envelope with $U = 0.20$ W/(m^2 K)

of carbon footprints in the following heating seasons. It can be noticed that already in the eight heating season the balance of all thermal insulations is negative, and the carbon footprint savings by reducing heat loss are greater than the carbon footprint due to the installation of insulation. Moreover, half of the analysed thermal insulation materials achieve this balance before the end of the third heating season.

The carbon footprint calculations of insulation materials demonstrated that the evaluation and consideration of their environmental impact per unit weight is inappropriate and can lead to misleading decisions, since it is imperative that the analysis considers also the difference in density (ρ) of each thermal insulation materials, as well as differences in their thermal conductivity (λ). Furthermore, the most effective measure to reduce the environmental impact of buildings is the correct choice of thermal insulation materials.

Comparison of carbon footprints of all fifteen analysed thermal insulation materials and other building materials, such as reinforced concrete, mortar or plaster, brick clay, window glass, PVC products, structural steel and aluminium, also supports the concept that a relatively small carbon footprint "investment" in thermal insulation materials yields significant savings in the operational time of a building. This holds true even if the building is to be thermally insulated well above the currently prescribed levels by the applicable regulation in most countries.

In addition to relatively low-carbon footprint, thermal insulation materials also contribute significantly to energy savings of buildings. Therefore, they belong to the very top of the most effective measures and investments to save energy and reduce indirect impacts on the environment, which has been proved by this analysis and confirmed by the results of other studies (Vattenfall 2015).

3.2 *Environmental Neutrality of a Building*

For the purpose of calculating environmental neutrality (the time needed for the carbon footprint due to the installation of thermal insulation to equal the carbon footprint of the difference of heat loss in the heating season between the current average external insulated building envelope and well-insulated building), the current average value of the thermal transmittance of the building envelope in Slovenia, $U = 0.52\ \mathrm{W/(m^2\ K)}$, was taken (STAT 2012). Thermal insulation thicknesses were calculated in the way that a thermal transmittance value of $0.20\ \mathrm{W/(m^2\ K)}$ of new, renovated external envelope is achieved.

The so-called environmental neutrality is achieved relatively soon for each of the analysed insulation materials. By thickening of the insulation layer, the carbon footprint due to the installation of thermal insulation increases, but the value of heat transfer U decreases, therefore reducing the carbon footprint of the heat loss through the building envelope.

From Table 5, where the requirement of thermal transmittance of the building envelope is the value of $U = 0.20\ \mathrm{W/(m^2\ K)}$, it can be concluded that the carbon footprint of the heat loss reduction due to the installation of thermal insulation

Table 5 Carbon footprint of built-in thermal insulation and the difference between the average insulated building envelope (carbon footprint of heat losses per heating season of 11.53 kg CO_2-eq./m^2) with a well-insulated building envelope (thermal transmittance $U = 0.20$ W/(m^2 K) with carbon footprint of heat losses per heating season of 4.43 kg CO_2-eq./m^2) per unit area (m^2) of the building envelope

Thermal insulation material	Carbon footprint per unit area (m^2) of thermal insulation for $U = 0.20$ W/(m^2K)	Carbon footprint of insulation for U-value 0.20 W/(m^2 K) and difference of heat losses through existing average external envelope and heat losses through well-insulated envelope ($U = 0.20$ W/(m^2 K)), per unit area (m^2) of external building envelope							
Heating season		1st	2nd	3rd	4th	5th	6th	7th	8th
Carbon footprint	kg CO_2–eq./m^2	kg CO_2–eq./m^2	kg CO_2–eq./m^2	kg CO_2–eq./m^2	kg CO_2–eq./m^2	kg CO_2–eq./m^2	kg CO_2–eq./m^2	kg CO_2–eq./m^2	kg CO_2–eq./m^2
EPS	9.47	2.38	**−4.72**	**−11.82**	**−18.91**	**−26.01**	**−33.10**	**−40.20**	**−47.30**
EPS with reflective additives	8.48	1.39	**−5.71**	**−12.80**	**−19.90**	**−27.00**	**−34.09**	**−41.19**	**−48.29**
XPS	33.74	26.64	19.54	12.45	5.35	**−1.74**	**−8.84**	**−15.94**	**−23.03**
PU polyurethane	25.73	18.63	11.53	4.44	**−2.66**	**−9.76**	**−16.85**	**−23.95**	**−31.04**
Glass wool—low density	5.47	**−1.62**	**−8.72**	**−15.81**	**−22.91**	**−30.01**	**−37.10**	**−44.20**	**−51.30**
Glass wool—high density	18.71	11.61	4.52	**−2.58**	**−9.67**	**−16.77**	**−23.87**	**−30.96**	**−38.06**
Rock wool—low density	14.32	7.22	0.12	**−6.97**	**−14.07**	**−21.16**	**−28.26**	**−35.36**	**−42.45**
Rock wool—high density	29.72	22.62	15.53	8.43	1.34	**−5.76**	**−12.86**	**−19.95**	**−27.05**
Wood fibre wool—low density	1.73	**−5.36**	**−12.46**	**−19.56**	**−26.65**	**−33.75**	**−40.85**	**−47.94**	**−55.04**
Wood fibre wool—high density	9.88	2.78	**−4.32**	**−11.41**	**−18.51**	**−25.60**	**−32.70**	**−39.80**	**−46.89**
Cellulose—recycled	4.56	**−2.53**	**−9.63**	**−16.73**	**−23.82**	**−30.92**	**−38.02**	**−45.11**	**−52.21**
Cork	43.56	36.46	29.36	22.27	15.17	8.08	0.98	**−6.12**	**−13.21**
Foam Glass	56.02	48.92	41.83	34.73	27.63	20.54	13.44	6.34	**−0.75**
Aerogel	47.33	40.23	33.13	26.04	18.94	11.84	4.75	**−2.35**	**−9.44**
VIP	53.52	46.43	39.33	32.23	25.14	18.04	10.94	3.85	**−3.25**

Negative bolded numbers indicate that the carbon footprint of installed thermal insulation is already covered by lower thermal energy losses and its carbon footprint throughout the building envelope is significant compared to the footprint of the initial "investment" in the insulation installation. After 7.89 heating seasons, the savings of environmental impacts are higher even than impacts caused by the installation of the thermal insulation with the highest environmental impact. The analysis showed that after eight heating seasons reduced environmental impact of buildings due to energy savings resulting from implementation of thermal insulation is higher than the impact of inserting insulation itself. More than one half (8 out of 15) of all of the evaluated thermal insulation types achieve environmental neutrality even before the end of the third heating season, and approximately three-fourths (11 out of 15) before the end of the sixth heating season. The smallest impact on the environment is caused by thermal insulation of wood–wool; the environmental neutrality is reached shortly after installation (0.57 heating seasons). This proves the extremely low-carbon footprint of installation of wood–wool insulation. Furthermore, thermal insulation materials have long life spans, during which the thermal conductivity does not deteriorate. Therefore, multiple returns of environmental impact caused by their installation in the building envelope are guaranteed.

It can be concluded that the environmental effects of thermal insulation materials in comparison with other building materials, which are embedded in the average building, are small. In addition, it should be noted that due to energy savings that thermal insulation materials provide in each heating season after their installation, they significantly contribute to the reduction of buildings' environmental impact. Therefore, thermal insulation should be ranked at the top of the most effective investments for energy saving and consequent reduction of environment impact of buildings.

4 Environmental Impacts of Buildings

Globally, infrastructure and building construction consumes 60 % of the raw materials extracted from the Earth (BREEAM 2014). From this volume, building represents 40, or 24 % of those global extractions. In the US, with only 4 % of the world's population resource consummation has reached a staggering 25 % of the total resources available in the world. A majority of these resources (60 % according to United States Green Building Council—USGBC) are consumed in the building industry. In Europe, the mineral extractions per capita intended for building amount to 4.8 tonnes per inhabitant per year, which is 64 times the average weight of a person, highlighting the need to work towards dematerialisation in building (Kutnar and Hill 2014).

Buildings, including their construction elements and materials, are extremely complex systems. They do not only undergo frequent changes, but also have multiple functions, are unique, contain many different elements, and have unclear system boundaries. A building uses most of its energy during its service life, which is about 90 % of the total life cycle energy (Citherlet and Defaux 2007; Newsham et al. 2009). Therefore, the adoption of the life cycle approach to design should not only account for current energy concerns, but should also integrate to long-term energy, environmental, and social impacts. Parallel to the development of sustainable building, the environmental impact of materials, products and new production processes throughout the life cycle of buildings [with an analysis of LCA and LCCA (Life Cycle Cost Assessment)] has to be assessed. Additionally, materials that are not harmful to human health and are recyclable should be used. Furthermore, for each developed product the sustainable use scenario at the end of its first life cycle should be determined. The proper choice of insulation can significantly reduce the environmental impact caused by the building's operational phase (service life). In the past few decades, several studies have been performed with the aim to reduce environmental impact of buildings. At the same time, new methods of assessing a building's sustainability are under development (Ding 2008; Haapio and Viitaniemi 2008; Sinha and Kutnar 2012). Newly published standards for sustainability for construction works (CEN 2012) open opportunities for EU-wide harmonization of calculations and reporting of a building's environmental impacts. The most important standards are EN 15804 for construction product Environmental Products Declarations (EPDs) and (CEN 2012) for assessment of environmental performance. In March 2011, the Construction Products Regulation (305/2011) was introduced, replacing the Construction Products Directive (2008/98/EEC). The Construction Products Regulation has come into full force as of July 2013. In order to develop a framework that allows for comparability of environmental performance between products, ISO 14025 (2009) was introduced. This describes the procedures required in order to produce Type III environmental declarations—the EPDs. This is based on the principle of developing product category rules (PCR), which specify how the information from an LCA is to be used to produce the EPD. For the construction sector, the core PCR is EN15804.

Since the buildings service life is often decades, eventually centuries, they are maintained and renovated during their life time period. ISO 15686-1 (2000) and 2 (2001) 'Building and constructed assets—service life planning'—Part 1 'General principles' and Part 2 'Service life predictions procedures', standard defines the service life of the building components according to their accessibility. Renovations have environmental impacts; if the building product is replaced before its predicted service life is finished, is the remaining service life wasted.

Different architectural, design and engineering solutions can achieve the same or similar function (bearing load, thermal or sound insulation, waterproof resistance, etc.) by using different material combinations in various constructional complexes in building designs, and hence, which is very important, with different primary energy uses and net CO_2 emission. The construction industry of tomorrow will favour materials and systems with low embodied energy that can easily be recycled and reused, based on sustainable supply chains. However, the influence of embodied energy on the life cycle balance is by many architects still not well understood. 'Zero waste' means maximum material efficiency and resource recovery, without any construction waste to landfill; it also means that buildings are fully demountable and fully recyclable at the end of their life cycle (Lehmann 2013).

For a product, the life cycle starts with procuring the raw material, primary processing, secondary processing or manufacturing, packaging, shipping and handling, installation, in-use energy consumption, maintenance, and end-of-life strategies. For buildings, the life cycle generally starts with the extraction of raw resources from the natural environment or recovery of materials from a previous use. The raw resources are then manufactured into useable products, such as steel, concrete, etc. The finished products are then shipped to the site, consuming energy in the process. On the site, the products are assembled into a building. During the service life of the building, it consumes energy. In due course, renovation or retrofit is performed on the building which uses materials and energy. Finally, the building is removed/demolished and its materials disposed of either as construction waste or recycled for reuse. Each of these steps consumes energy and materials, and produces waste. With LCA analysis it is possible to quantify how a building product or system affects the environment during each phase of its life (Berry et al. 2014). Possible reductions of emissions could be mainly as a result of using wood or wood products instead of concrete, bricks, steel or aluminium, all of which require much more process energy than wood or other bioproducts (Buchanan and Bry Levineb 1999). Natural fibres are recyclable, biodegradable, inexpensive, and abundant, unlike common synthetic reinforcing fibres, as glass, carbon, etc. Timber as a natural raw material requires minimal energy input into the process of its becoming construction material. Timber represents one of the best choices for energy-efficient construction, since it also functions as a material with good thermal transmittance properties, if compared to other construction materials. Moreover, timber has good mechanical properties and ensures a comfortable indoor climate in addition to playing an important role in the reduction of CO_2 emissions (Premrov and Žegarac-Leskovar 2013).

Designing timber buildings requires more careful detailing and precise planning than other construction methods. Generally, condensation can occur where moist air comes into contact with a surface of a lower temperature. Air always contains water vapour in varying quantities; its capacity to do so is related to its temperature—warm air holds more moisture than cold air. When moist air comes into contact with colder air or a colder surface (e.g. a timber element), the air is unable to retain the same amount of moisture and the water is released to form condensation in the element. The moisture from condensed water causes timber to decay, as the damp causes wet rot inside the walls. This is often hard to detect and may not be noticed until mould growth or rotting of material actually occurs (Lehmann 2013).

Kitek Kuzman and Kutnar (2014) described most common timber construction systems, which are: timber panel construction, timber frame construction, and solid wood construction systems. In solid timber construction, we differentiate between two types of buildings: buildings with solid log walls and buildings with walls made from adhesively bonded wood elements. The walls are usually clad with insulation and panels on one side. However, it is also possible to use no cladding, leaving beams or engineered wood panels visible. The external walls may be protected by wooden cladding, roughcast, or brick facade. Ceiling structures are most often made from solid wood glued panels or ceiling beams, panelled and insulated on both sides. Recently, the CLT system has been a focus of interest in the timber house construction industry. Research in the field related to CLT buildings has so far frequently been part of a broader focus on examining the low-carbon properties of timber buildings, their structural performance, their fire performance and how forest and wood products, including recent innovations in engineered wood products, may be used more widely in the construction of multi-storey buildings. A major focus of timber building research compares the environmental impacts of timber with that of steel, concrete and masonry. Wood-framed construction requires less energy, and emits less CO_2 to the atmosphere, than concrete-framed construction. The life cycle emissions difference between the wood- and concrete-framed buildings ranges from 30 to 130 kg C per m^2 of floor area (Gustavsson et al. 2006). The carbon mitigation efficiency, expressed in terms of biomass used per unit of reduced carbon emission, is considerably better if the wood is used to replace concrete building material than if wood is used directly a biofuel (Gustavsson et al. 2006). A 55 % reduction in GHG emissions (carbon dioxide) is reported to be achievable by using timber, compared to steel and concrete, confirming that timber provides a lower carbon solution. For multi-storey buildings specifically, Gustavsson et al. (2006) compared two functionally equivalent four storey buildings, one framed in timber, the other framed in reinforced concrete, finding that the timber framed building used 28 % less primary energy for materials and emitted 45 % less carbon than in the concrete-framed building (Lehmann 2013).

Wood is exceptional compared with other materials in terms of stored carbon and emissions of carbon dioxide from fossil fuel energy used in manufacturing. The possible reduction in emissions is mainly a result of using wood in place of brick and aluminium, and to a lesser extent steel and concrete, all of which require much more process energy than wood. Buchanan and Honey (1994) showed that the energy required to manufacture a typical house is of a similar order of magnitude to the energy required to heat the house over a 25-year period. Buildings with a large timber content have much lower embodied energy and lower resulting emissions.

The use of prefabricated modular construction made of solid wood, such as load-bearing panels, CLT panels and bio construction products with the possibility of their disassembly, offers, among other benefits, significant opportunities for greenhouse gas (GHG) emission reduction and waste avoidance. Factory prefabrication dramatically reduces construction time on-site. Prefabricated elements are not only extremely suitable for implementing Building Information Modelling (BIM), but also their technical and architectural design is efficient, simple and fast. Furthermore, wood buildings made of cross-laminated wood walls have a substantially lower life cycle carbon emission than a concrete buildings, even if the carbon benefits of end-of-life concrete management is included, before (as a row material) or after (as an ideal recycle process) construction.

The concrete- and steel-framed housed use 2.5 and 2.8 times more non-bioenergy, respectively. CO_2 emission is lower for the wood-framed houses due to carbon storage in wood products and reduces use of fossil fuels (Lippke et al. 2004). Increasing the use of wood material in construction is a potential option for reducing net CO_2 emission because of the relatively low energy needed to manufacture wood products compared with alternative materials, the storage of carbon in wood building materials, and the increased availability of biofuels from wood by-products (Gustavsson et al. 2006). Timber stores 0.8 tonnes of CO_2 within 1 m^3. In comparison, the production of concrete, aluminium and steel are one-way energy intensive processes that release large amounts of CO_2 into the atmosphere.

Studies show that a 17 % increase in wood usage in the New Zealand building industry could result in a 20 % reduction in carbon emissions from the manufacture of all building materials, being a reduction of about 1.5 % of New Zealand's total emissions (Buchanan and Bry Levineb 1999). The reduction in emissions is mainly a result of using wood in place of brick and aluminium, and to a lesser extent steel and concrete, all of which require much more process energy than wood. Hence there is a net flux of carbon to the atmosphere from the forest sector. Besides this Buchanan (1999) lists the following facts about carbon stored in the buildings

- A change in the mix of construction materials, with increasing emphasis on wood as a building material, could have significant implications for global forestry, global energy requirements and global carbon dioxide emissions.
- Storage of carbon in wood products is very small in relation to total carbon emissions and cannot be used to offset emissions in the long term.
- Carbon emissions from the manufacturing of construction materials are far more important than the carbon stored in the materials. For example, if the use of wood in housing is increased at the expense of energy intensive materials, the emission reduction can be up to fifteen times the amount of carbon stored in the extra wood products.
- Given the finite life of all wood products, total storage of carbon in wood products remains almost constant over time, whereas the cumulative carbon emissions from manufacturing of wood products continue to increase. At current rates, the cumulative carbon emissions from fossil fuels over a 150-year period approximately equal the stored carbon.
- As a corollary, a shift from wood to non-wood materials would result in an increase in energy requirements and atmospheric carbon emissions.
- Considering global carbon emissions, wood is a much better building material than more energy intensive materials such as bricks, aluminium, steel and concrete, but an increase in wood use can only be justified if there is a corresponding increase in the area of forest available for long-term sustainable management.

Gustavsson et al. (2006) calculated carbon dioxide (CO_2) and methane (CH) emissions from the construction of a multi-storey building, with either a wood or a concrete frame from life cycle and forest land-use perspectives The primary energy input (mainly fossil fuels) in the production of building materials was found to be about 60–80 % higher when concrete frames were considered instead of wood frames.

4.1 Durability of Timber Construction

Durability is defined as the ability of a building or any of its components to perform the required functions in a service environment over a period of time without unforeseen cost for maintenance or repair. Wood is durable material, which has to be accompanied with appropriate building applications and design. The natural durability of wood has been proven by the multitude of buildings that have stood for centuries. While wood's natural attributes make it a sustainable building material, they also make wood vulnerable to decay and wood destroying insects. Proper design, installation, and detailing are critical to ensure long-term durability. When wood is used in exposed applications, or in areas where it is subjected to moisture and insects, it must be protected with mechanical barriers, coatings and, in some instances, preservative treatments. In sustainable design, "durability" is also increasingly being included on priority lists under the assumption that designing for longevity is an environmental imperative. However, this is unsupported in the absence of LCA and accurate lifespan predictions. In the worst case, designing for longevity can lead to design choices that are well intentioned but, in fact, yield poor environmental results. Rather than attempt to predict the future and design permanent structures with an infinite lifespan, design for easy adaptation, and material recovery should be considered (Kitek Kuzman and Kutnar 2014).

The durability of houses depends on material, system type, application, and hazard exposure. Causes of degradation can be placed into three groups: physical processes (e.g. freezing, thawing, surface wear, cracking, etc.); chemical processes (e.g. corrosion, alkali–silica reaction, sulphate attack, etc.) and biological processes (e.g. fungi/rot, marine borers, termites, etc.). Durability, which assures the longevity of a house, can be enhanced by proper design and construction, by preservative treatments, through proper use of finishes and by properly controlling interior moisture (Moody and Sherwood 1986). Computer software tools for modelling all types of moisture flows throughout entire buildings have been developed. The end of a buildings service life may be due to technical, economical, environmental reasons or planning changes, aesthetic and societal reflections. The key to improved reliability of timber in construction is the availability of prediction models for service life performance of timber products and components (Foliente et al. 2002). For development of these prediction models the use of full-building specimens is essential for obtaining real-time data on houses durability. The reasons for limited tests of full-scale houses lie not only with the prohibitive cost of full-scale tests, but also with the unique properties of wood (variability, nonhomogeneous mechanical properties in three orthogonal directions, load capacity that is affected by duration of loading). Development of a degradation model for wood construction has not received as much research attention as models for concrete and steel construction. Foliente et al. (2002) provided an overview of different approaches to durability design and described key aspects of the Australian approach to developing an engineered method of durability design for wood construction. The study concluded that probabilistic prediction models of the effects of various durability hazards on the long-term performance of wood are possible. However, these models are largely based on a mixture of data obtained from field tests on small clear pieces of wood and not on actual buildings (Kitek Kuzman and Kutnar 2014). Timber is an organic material and therefore combustible; however, despite its combustibility, timber particularly in larger sections performs better in a fire than the equivalent sections of exposed steel or aluminium (Lehmann 2013). To protect timber from insect attack, a wide range of effective preservatives are commercially available for use under controlled conditions (Lehmann 2013).

4.2 Reuse, Upgrading, and Recycling of Construction Elements After Service Life of a Building

The reuse of building materials after demolition is estimated to increase in the future. Today, up to 90 % of demolition wood is deposited in landfills in Sweden, but it is assumed that new, economic and legislative measures will decrease deposition and increase reuse (Johansson 1995). Examples of wood products which could be reused are timber previously used for structural purposes, and non-structural wood product s such as floors, doors, window frames, stairs, etc. However, it is not possible to reuse all the wood products in a building, due to technical and economic constraints. It can be assumed that half of the wood construction materials can be reused in a new building, leading to a decrease in energy use in the production of new building materials of 50 %, while the other half is used to replace fossil fuels. This estimate is uncertain and will depend on several factors, such as the construction of the building, how demolition is carried out, and the quality of the wood products (Swedish Board of Housing, Building and Planning 1998). To achieve a more reliable estimate, more research is required in this field (Gustavsson et al. 2006).

A minor fraction of the demolition wood may be left at the demolition site as all the demolition wood cannot be collected for technical reasons. Some of the demolition wood may also be unsuitable for reuse or as a fuel, due, for example, to chemical treatment. Wood products damaged by fungus or insects cannot be reused and the heating value of such wood is much lower than that of healthy timber. These aspects have to be considered as well. They may have a minor impact on the results, especially if building techniques are adapted to increase the reuse of building materials and the use of demolition wood for energy purposes (Gustavsson et al. 2006).

Recovered woody material after demolition or disassembly of the building can be either burned as biofuel, use as a landfill or better used as input for further processing into other wood or wood-based products. Such reprocessing of wood materials or constructions at the end of the building life cycle can have significant effects on the energy and carbon (GWP and GHG) balances of the materials, constructions or whole buildings (Gustavsson et al. 2006).

4.3 Indoor Air Quality

In the past, outdoor air pollution received the majority of society attention, while indoor air quality just recently became a major concern for human health. Indoor air pollutants can be generated inside the home through human activities (smoking, use of oven cleaners, disinfectants, carpet shampoos, insecticides, paints) or the pollution sources can be materials that went into the construction of building itself (particleboard furniture, cabinets, flooring, isolation materials, etc.). Despite the strict standards of manufacturing and use, the emission of VOCs is still an issue and becomes even more important in buildings fulfilling modern standards of heat insulation. Due to the reduced air exchange rates within low energy and plus energy buildings, certain volatile chemicals present in interior equipment or resulting from activities of human beings tend to accumulate in the building. Oanh and Hung (2004) and Friedman et al. (Friedman et al. 2009) listed common indoor hazards and provided their description, primary household sources, health effects, typical exposure levels and steps to reduce exposure. Major household pollutants are radon, formaldehyde, pollutants of biological origin (e.g. bacteria, moulds, mildew, viruses, animal dander, dust mites, pollen, etc.), volatile organic compounds (VOCs), pesticides, lead and asbestos. Furthermore, moisture can be listed as an air contaminant, while it has high effect on indoor air quality by influencing the emission of some other pollutants, e.g. formaldehyde. Norbäck and Nordström (2008) discussed the indoor air quality and its effects on sick building syndrome (SBS), situations in which building occupants experience acute health and comfort effects that appear to be linked to time spent in a building, but no specific illness or cause can be identified. Among all of indoor air pollutants formaldehyde is one of the most studied chemicals in use today. Formaldehyde, also known as formalin, embalming fluid, is a reactive colourless gas with a strong pungent odour at elevated concentrations and an excellent preservative, disinfectant, and bonding agent. Formaldehyde is currently used in thousands of products as an adhesive, bonding agent and solvent (Minnesota Department of Health Fact Sheet 2010).

Formaldehyde is an important industrial chemical used to manufacture building materials and to produce many commercial, industrial and household products (DeVany 2007). Formaldehyde-based chemicals are used in pressed wood products, urea-formaldehyde foam insulation, embalming fluids, carpets, combustion appliance, clothing, and tobacco (Heimlich 2008). Many governmental bodies have placed limitations on formaldehyde emissions from products; an overview of these values from European and worldwide bodies is provided by Salthammer et al. (2010). Although emissions from products are often regulated, small concentrations of formaldehyde are a normal part of our environment. It is present in synthetic fabrics, shampoos, and cosmetics (Minnesota Department of Health Fact Sheet 2010). Furthermore, due to its importance in various metabolic processes, formaldehyde is naturally present in human body with concentrations of approximately one to two parts per million (ppm) in blood (Formaldehyde Council 2007). The International Agency for Research on Cancer (IARC) classified formaldehyde as a "probable human carcinogen" in 1995, meaning that it causes cancer in animals and likely humans, and in 2004 reclassified it as a "human carcinogen" (DeVany 2007). The California Air Resource Board (California Air Resource Board 2004) has stated the concentrations of formaldehyde that have health effects. High concentrations of formaldehyde can cause occupational asthma (Kim et al. 2001). Formaldehyde concentrations of 50–500 parts per billion (ppb) cause odour, eye and noise irritation, nasal stuffiness, and lung discomfort. Variable concentrations cause allergic reactions, and worsening of asthma symptoms, while the level causing cancer is not known. When the IARC classified formaldehyde as known "human carcinogen", the California Air Resources Board (CARB) voted in 2007 to adopt the nation's most stringent regulations on formaldehyde emissions from particleboards and other composite wood products, with the purpose of improving the indoor air quality across the United States (Lent 2007). Furthermore, this decision attested to the potential power of the LEED rating systems to dramatically accelerate the transformation of the building materials market toward healthier materials (Lent 2007). Additionally, the first formaldehyde emission standards for composite wood products were implemented in January 2009 (California Air Resource Board, 2004). Of all home products containing formaldehyde, pressed wood (panelling, particleboard, hardwood plywood, medium density fibreboard) has the highest concentrations (Heimlich 2008). The manufactures were forced to reduce levels of outgassing or off gassing (gradual release of formaldehyde from the parent material) in structural panel products, which include softwood plywood, OSB, and composite panels. However, there are numerous studies still reporting formaldehyde contamination of residential and non-residential indoor environments, especially in new homes, in which the air change rate necessary to maintain formaldehyde compliance is higher (Gilbert et al. 2008; Mantanis et al. 2007; Maruo et al. 2007; Offermann et al. 2007). Formaldehyde levels in the air below 0.1 ppm are considered safe (Oanh and Hung 2004; Kitek Kuzman and Kutnar 2014).

References

Berglund L, Rowell RM (2005) Handbook of wood chemistry and wood composites. CRC Press, Boca Raton, pp 279–301

Berry S, Davidson K, Saman W (2014) Defining zero carbon and zero energy homes from a performance-based regulatory perspective. Energy Effi 7:303–322

Bodig J, Jayne B (1982) Mechanics of wood and wood composites. Van Nostrand Reinhold Company, New York 712

BREEAM (2014) www.breeam.org. Accessed 1 May 2014

Buchanan AH (2006) Can timber buildings help reduce global CO_2 emissions? In: Proceedings of the world conference on timber engineering. Portland, Oregon, 6–10 Aug

Buchanan AH, Bry Levineb S (1999) Wood-based building materials and atmospheric carbon emissions. In: Environmental science and policy. University of Canterbury, Christchurch, pp 427–437

Buchanan AH, Honey BG (1994) Energy and carbon dioxide im-plications of building construction. Energy Build 20:205–217

California Air Resource Board (CARB) (2004) Formaldehyde in the home. http://www.arb.cagov/research/indoor/formaldGL08-04.pdf. Accessed 3 Apr 2014

CEN/TC 350 (2012) Sustainability of construction works

Citherlet S, Defaux T (2007) Energy and environmental comparison of three variants of a single family house during its whole life span. Build Environ 42:591–598

Climate Change (2001) IPCC third assessment report. The scientific basis. http://www.grida.no/climate/ipcc_tar/. Accessed 03 Mar 2010

Dehne M, Krueger U (2006) Fire protection in multi-story timber construction, detail. no. 10/2006, pp 1142–1144

DeVany MC (2007) The serious public health issues resulting from formaldehyde exposures within FEMA travel trailers issued to hurricane disaster victims, and recommended action items. Testimony before committee on oversight and government reform U.S: house of representatives. http://s3.amazonaws.com/propublica/assets/formaldehyde/devany_testimony_070719.pdf. Accessed 3 Apr 2014

Ding CKC (2008) Sustainable construction: the role of environmental assessment tools. J Environ Manage 86:451–464

Dinwoodie JM (2000) Timber: its nature and behaviour, 2nd edn. EFN Spon, London 257 p

Ecoinvent 3.0 (2013) Swiss centre for life cycle inventories. Dübendorf, Switzerland

European Committee for Standardisation (CEN) (2012) EN 15804: sustainability of construction works—Environmental product declarations—core rules for the product category of construction products

European Parliament, Council (2008) Directive 2008/98/EC of the European parliament and of the council of 19 November 2008 on waste and repealing certain directives. Directive. Brussels: European parliament European parliament. http://eur-lex.europa.eu/LexUriServ/LexUriServ.do?uri=CELEX:32008L0098:EN:NOT. Accessed 15 Sept 2013

Foliente G, Leicester B, Wang C, Mackenzie C, Cole I (2002) Durability design for wood construction. Forest Prod J 52(1):10–19

Formaldehyde Council, Inc. (2007) Formaldehyde and human health. http://www.formaldehyde.org/newsroom/documents/FormaldehydeandHumanHealth.pdf. Accessed 15 June 2014

Friedman D, Bliss S, Wiley S (2009) Guide to finding and reducing common indoor air pollutants and contamination sources. Inspect a pedia. http://www.inspect-ny.com/BestPractices/IAQ_Hazard_Table1.htm. Accessed 15 June 2014

Gilbert NL, Guay M, Gauvin D, Dietz RN, Chan CC, Levesque B (2008) Air change rate and concentration of formaldehyde in residential indoor air. Atmos Environ 42(10):2424–2428

Gustavsson L, Pingoud K, Sathre R (2006) Carbon dioxide balance of wood substitution: comparing concrete- and wood-framed buildings, Springer, Berlin. Greenhouse gas balances in building construction: wood versus concrete from life-cycle and forest land-use perspectives. Energy Policy. Environmental and energy systems studies. Lund University, Lund Institute of Technology, Lund

Haapio A, Viitaniemi P (2008) A critical review of building environmental assessment tools. Environ Impact Asses 28:469–482

Heimlich JE (2008) Formaldehyde. The invisible environment fact sheet series. http://ohioline.osu.edu/cd-fact/pdf/0198.pdf. Accessed 20 June 2014

ISO 14040 (2006) Environmental management—life cycle assessment—requirements and guidelines. Standard. Geneva, Switzerland

ISO 14025 (2009) Environmental labels and declarations—type III environmental declarations—principles and procedures, standard. International Standards Organisation, Geneva

Johansson B (1995), Bygg- och Rivningsmateriali Kretsloppet (Recycling of building and demolition materials). Report 17:1995, Swedish Building Research Council, Stockholm, Sweden

Karjalainen T, Kellomaki S, Pussinen A (1994) Role of wood-based products in absorbing atmospheric carbon. Silva Fennica 28(2):67–80

Karjalainen T, Pussinen A, Kellomaki S, Makipaa R (1999) Scenarios for the carbon balance of finnish forests and wood products. Environ Sci Policy 2(2):165–175

Kim CW, Song JS, Ahu YS, Park SH, Park JW, Noh JH, Hong CS (2001) Occupational asthma due to formaldehyde. Yonsei Med J 42(4):440–445

Kitek Kuzman M, Kutnar A (2014) Contemporary Slovenian timber architecture for sustainability. Green energy and technology. Springer International Publishing Switzerland. ISSN 1865-3529

Košir M, Krainer A, Dovjak M, Perdan R, Kristl Ž (2010) Alternative to the conventional heating and cooling systems in public buildings. J Mech Eng 56(9):575–583

Kunič R (2007) Planning an assessment of the impact of accelerated ageing of bituminous sheets on constructional complexes: *doctoral thesis*. University of Ljubljana, Faculty of Civil and Geodetic Engineering, Ljubljana

Kunič R, Krainer A (2009) Economical thickness of thermal insulation layers in ETICS façade systems. Gradbeni vestnik 58(2): 306–311

Kunič R, Krainer A (2010) Economical thickness of thermal insulation layers in flat roofs. Gradbeni vestnik 59(1): 6–12

Kutnar A, Hill C (2014) Assessment of carbon footprinting in the wood industry. In: Muthu SS (ed) Assessment of carbon footprint in different industrial sectors, vol 2, (EcoProduction). Singapore (etc.), Springer, cop. 2014, pp 135–172

Lehmann S (2013) Low carbon construction systems using prefabricated engineered solid wood panels for urban infill to significantly reduce greenhouse gas emissions. Sustain Cities Soc J. Research Centre, University of South Australia, Adelaide, Australia

Lent T (2007) LEEDing on formaldehyde in California. Healthy building news. http://www.healthybuilding.net/news/070511formaldehyde.html. Accessed 20 June 2014

Lippke B, Wilson J, Perez-Garcia J, Boyer J, Meil J (2004) CORRIM: life-cycle environmental performance of renewable building materials. For Prod J 54(6):8–19

Maclaren JP, Wakelin SJ (1991) Forestry and forest products as a carbon sink in New Zealand. FRI Bull No 162. Ministry of Forestry, Rotorua, New Zealand

Mantanis G, Vouli E, Gonitsioti C, Ntalos G (2007) Formaldehyde in indoor air of new apartments in Greece. COST action E49 conference, measurement and control of VOC emissions from wood-based panels. http://users.teilar.gr/~mantanis/Formaldehyde.pdf. Accessed 20 June 2014

Maruo YY, Yamada T, Nakamura J (2007) Sensor element for indoor formaldehyde measurement. NNT technical review. https://www.ntt-review.jp/archive/ntttechnical.php?contents=ntr200910sf5.html. Accessed 20 June 2014

Minnesota Department of Health Fact Sheet (2010) Formaldehyde facts. http://www.health.state.mn.us/divs/eh/indoorair/voc/formaldehyde.pdf. Accessed 20 June 2014

Moody RC, Sherwood GE (1986) Light-framed construction research at USDA forest products laboratory—a status report. Appl Eng Agric 2(2):167–173

Newsham GR, Mancini S, Birt BJ (2009) Do LEED-certified buildings save energy? Yes, but. Energy Build 41:897–905

Norbäck D, Nordström K (2008) Sick building syndrome in relation to air exchange rate, CO_2, room temperature and relative air humidity in university computer classrooms: an experimental study. Int Arch Occup Environ Health 82:21–30

Oanh NTK, Hung YT (2004) Indoor air pollution control. In: Wang LK, Pereira NC, Hung YT (eds) Handbook of environmental engineering, vol 2: advanced air and noise pollution control

Offermann F, Robertson J, Brennan S, Springer D (2007) Window usage, ventilation, and formaldehyde concentrations in new California homes: summer field sessions. http://www.ieesf.com/pdf/SummerFieldResults.pdf. Accessed 20 June 2014

PAS 2050 (2011) Specification for the assessment of the life cycle greenhouse gas emissions of goods and services

Pfundstein M et al (2008) Insulating materials. Institut für Internationale Architektur-Dokumentation, Munich

Premrov M, Žegarac-Leskovar V (2013) Energy-efficient timber-glass houses. Green energy and technology. Springer, London, 178 p

PURES (2010) Rules on efficient use of energy in buildings. Off J Rep Slovenia, No. 52/2010, XX, 7840–7847

Salthammer T, Mentese S, Marutzky R (2010) Formaldehyde in the indoor environment. Chem Rev 110(4):2536–2572

Schlamadinger B, Marland G (1996) The role of forest and bioenergy strategies in the global carbon cycle. Biomass Bioenergy 10:275–300

Simapro (2009). SimaPro analyst indefinite, product ecology consultants. PEC, The Netherlands, www.pre.nl/default.htm

Sinha A, Kutnar A (2012) Carbon footprint versus performance of aluminium, plastic and wood window frames from cradle to gate. Build (Basel) 2:542–553

STAT (2012) www.stat.si. Statistical office of the Republic of Slovenia. Accessed 25 June 2012

Vattenfall (2015) www.vattenfall.com. Accessed 25 Jan 2015

Wegener G, Pahler A, Tratzmiller M (2010) Bauen mit Holz = aktiver Klimaschutz. Ein Leitfaden, Munich, Holzforschung Muenchen

Winjum JK, Brown S, Schlamadinger B (1998) Forest harvests and wood products: sources and sinks of atmospheric carbon dioxide. For Sci 44(2):272–284

Zabalza I, Aranda A, Scarpellini S, Diaz S (2009) Life cycle assessment in building sector: state of the art and assessment of environmental impact for building materials. In: 1st international energy, life cycle assessment and sustainability workshop and symposium (ELCAS), Nisyros, Greece

Business—Bioproducts in the Bioeconomy

Anders Roos

Abstract This chapter considers the key issues of, and challenges for, the business development of forest-based bioproducts, defining the conditions and improvement potential for different product types: established forest products (building materials, paper and wood energy), potential large-volume bioproducts (e.g. liquid biofuels), and high value-added products (biomaterials, new chemicals, and pharmaceuticals). These products' respective competitive position, including success factors, are analyzed together with their innovation and market development potential, barriers, and production economics. We illustrate that key challenges differ between "old" bioproducts such as lumber, where progress is connected with design development, and the incremental improvement of industrial processes, and "radical" innovations of new materials and substances, which involve new market development. The economic conditions for the production and marketing of bioproducts are discussed. The chapter describes the innovation process for forest-based products and challenges to a successful market launch, using examples illustrating the range of potential bioproducts.

Keywords Bioeconomy · Business development · Innovation · Marketing · Competitiveness

1 The Bioeconomy and the Forest Business Community

The forest sector plays an important role in the global economy because of the range of products, services and innovation potential it hosts, and it is crucial for sustainable development because it affects many people's incomes and livelihoods. However, this role relies on the sector's ability to provide value for both producers and users in competitive markets, and on the its capacity to innovate and improve.

A. Roos (✉)
Department of Forest Products, Swedish University of Agricultural Sciences, Uppsala, Sweden
e-mail: anders.roos@slu.se

A. Kutnar and S.S. Muthu (eds.), *Environmental Impacts of Traditional and Innovative Forest-based Bioproducts*, Environmental Footprints and Eco-design of Products and Processes, DOI 10.1007/978-981-10-0655-5_6

Therefore, the forest-based bioeconomy rests on the capabilities of the sector's actors—authorities, forest owners, the forest industry, innovators, researchers, forest communities and customers of forest products—and on how they interact.

The forest sector is founded on established and mature industries, producing mainly lumber, panels, pulp, and paper. Forests also generate energy in different applications, for domestic use, in industrial processes, and in refined forms. Although these traditional uses dominate in terms of quantity and economic significance, the current discourse about forests in the bioeconomy also notes the sector's new and improved product offerings, transportations fuels, materials, chemicals, and pharmaceuticals.

The breadth of the business opportunities in forest products is not a completely new notion, highlighted, for instance, by Glesinger (1949), who stated that much of current practices based on petrochemicals can be achieved with wood. On the same note, Goldstein (1978) indicated that the future potential of wooden raw materials lies in their potential for chemical production; later, Enriquez-Cabot (1998) referred to the range of opportunities in biorefinery processing for building new substances, such as chemicals, cosmetics, pharmaceuticals, foodstuffs, and energy. Hardy (2002) perceived all business opportunities for these substances in the agricultural sector, and Duchesne and Wetzel (2003) saw similar opportunities in the forest industries.

In the wake of the economic problems in the beginning of the new millennium and increased environmental concerns, the forest sector is considered by some stakeholders to be suitable for a particular role in the bioeconomy as a provider of economic development, jobs, innovation, and services (c.f. Swedish Forest Industries Federation 2012; Forest Industries Finland 2013; Puddister et al. 2011). The forest industry claims that increasingly enhanced forest management methods and the upgraded production of both established and new bioproducts can boost its contribution to sustainable development (UNEP 2011, p. 21). The forest's role in the bioeconomy is therefore clearly stated in the sector's vision document, which indicates that the "forest-based sector is a key actor in and enabler of the bio-based society" (FTP 2015). The EU's forest strategy highlights the forest's multiple uses, as well as its potential to create a competitive forest sector in the bioeconomy, underlining the importance of building with wood, and considering innovation and research activities in the whole forestry-wood chain (EC 2013). The envisioned strategies, tasks and measures have influenced both public funding for research, development, and innovation, and the industrial community's readiness and receptiveness to the concept. These aspirations and expectations concerning a forest-led bio-based economy are reflected in the relevant research agenda (e.g., in EU-Horizon 2020 2015; FTP 2015).

The expectation that bioproducts will replace oil-based fuels, products and compounds is currently cheered by policymakers for enabling sustainable economic growth. Bioproducts can meet a growing market demand, it is said, while safeguarding the capacity of natural capital, resource stocks, land and ecosystems. Moreover, bioproducts should promote economic activity fueled by research and innovation in the biological sciences and combine employment and sustainable global development (OECD 2011; White house 2012; EC 2012; United Nations 2012). Different discourses do, however, reflect opposing perceptions of the bioeconomy, ranging from a technocratic and instrumental approach to ecological modernization and sustainable development discourses (Pülzl et al. 2014). As Pfau et al. found, expectations about the bioeconomy also differ; they are seen by some as inherently sustainable and by others as being a new brand for old, unsustainable forest industry practices. The sector's role is questioned mainly by environmental organizations, and the industry's true objectives are not entirely trusted (Hall et al. 2012).

By definition, the bioeconomy refers to an economy that substitutes bio-based, renewable materials and services for fossil resources. This role of biomass, including wood, in the production of chemicals, energy carriers, structural building and other materials is frequently highlighted in the Intergovernmental Panel on Climate Change (IPCC) reports on possible avenues to mitigate climate change (IPCC 2014, Chaps. 9 and 11). The report lists bioenergy use as an approach to substitute biofuels for fossil fuels in an effort to mitigate climate change. In order to achieve the high potential deployment levels of biomass for energy, increases in competing fiber demand must be moderate, land must be properly managed, and forestry yields must increase substantially. Expansion of bioenergy in the absence of the monitoring and good governance of land use carries the risk of low greenhouse gas (GHG) benefits and the potential for competing demands for virgin fiber by the forest industry and the biomass energy sector. Burning virgin fiber may not be better than using it in long-life products and incinerating it at the end of life. However, the introduction of restrictions on raw material use could, on the other hand, lead to economic welfare losses and misallocations. The optimal approach requires appropriate analysis.

However, the term bioeconomy also denotes the development of completely new products. In official documents, the bioeconomy is—or is expected to be—driven by the improvement of existing products and the innovation of new, superior products in a competitive market. These aspirations have gradually been reflected in research and in corporate strategies.

The forest-based bioeconomy concept is here seen as constituted by three main criteria: forest resources replacing raw fossil materials; that this process is sustainable; and that it relies on innovations and entrepreneurship in competitive markets. Figure 1 highlights that some ideas presented under the bioeconomy label do not meet all these criteria. For instance, an innovative forest product may rely on unsustainable forestry methods. Hence, the fulfillment of all three requirements must be specified, explained, and ultimately confirmed by facts on a case-by-case basis. However, in this chapter, we focus on the competiveness circle because to be viable any sustainable alternative has to provide better value for money to make a significant difference in the market.

Several factors influence the success of new forest-based products—and there are many reasons why excellent ideas can be stalled. Although policy and research documents on bioproducts offer interesting aspects regarding the development and market introduction of new offerings, they frequently do not describe these hindrances to market success both for ideas that involve incremental improvements in existing forest product groups and for the innovation of completely new

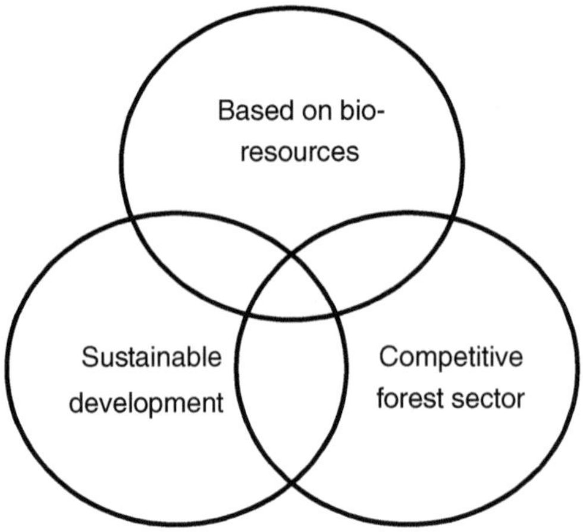

Fig. 1 Three components of the bioeconomy

bioproducts that could replace incumbent fossil-based alternatives. Furthermore, the business aspects of a forest-based bioeconomy involve the management of the innovation process, based on thorough knowledge of customer needs, and the technological challenges that sector actors face. The innovation and development process must therefore use a mix of resources, select the optimal solutions and collaborators to turn the idea into product concepts, and, finally, fine-tune the production and distribution chain. This process, of course, implies optimal business networks and a deep understanding of customers' needs. Furthermore, these activities take place in a legal, policy, normative and technology framework.

Forest products, including biorefinery outputs that are produced and used in large bulk quantities and with somewhat lower margins are distinct from increasingly differentiated products with more narrow customer bases, as well as from various product types in between these extremes. The former group is similar to traditional forest industry products, whereas differentiated products with short life cycles may become more important in a future, customer-focused bioeconomy (Fig. 2).

The development of new innovative forest-based products should likely be based on existing forest products and technologies. More specifically, new bioproducts often originate from by-products in the pulp and paper industry, and many actors currently entering the production of new bio-based products were formerly traditional forest enterprises. The new forest-based bioeconomy will also rest on existing sunk investments in infrastructure and facilities. Consequently, the traditional forest industries constitute the basis for the further development of new bioeconomy offerings (although they do not always fulfill all three criteria in Fig. 1), providing the sourcing infrastructure and network, know how and facilities. This argument motivates the focus and outline of this chapter. First, traditional pulp and paper, lumber and panels are presented together with their products, business realities and business strategies. Outputs emblematic of a bioeconomy are then described, bioenergy, and materials and chemicals. Thereafter, and based on the three

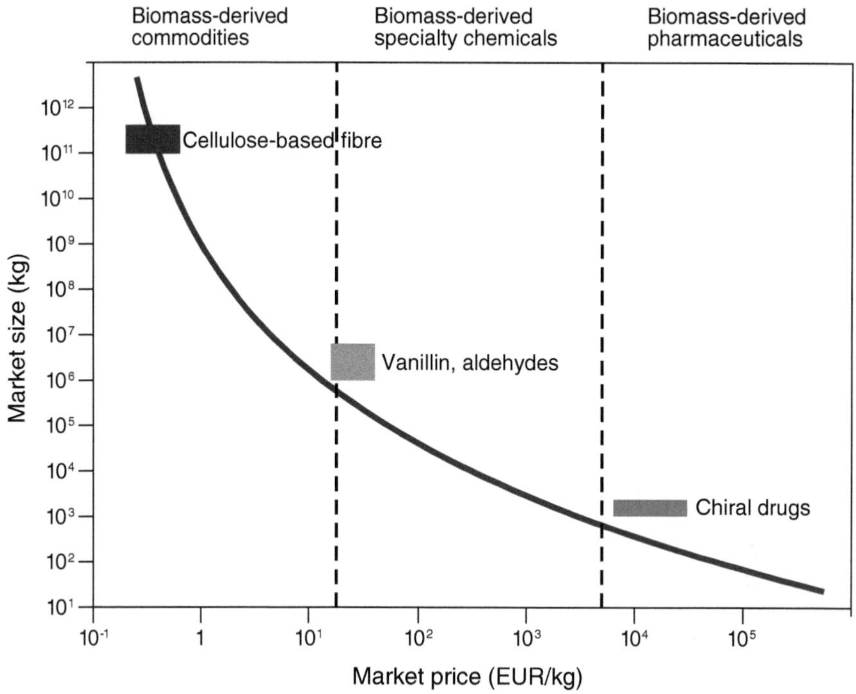

Fig. 2 Market price-size relationships for bio-based products. *Source* Sandén and Pettersson (2014)

categories (Fig. 1), key issues and considerations for business success are identified. This section also illustrates the approach by presenting a method for justly assessing the economic effects of different product mixes in biorefineries. A synthesis and conclusions section is found at the end of the chapter.

2 Harvesting and Established Forest Products

The forest products industry will likely be the basis for the introduction of new and innovative forest-based products. Both established and new forest-based products are processed jointly at several stages in the supply chain before differentiation and distribution. In silviculture, logging and transport are examples of joint processes. Moreover, even newer and more innovative products and chemicals will be produced in adapted forest enterprises, jointly with regular products; and oftentimes, investment capital in new production technologies will have to be earned in the traditional forest sector. Naturally, capital investments are scrutinized carefully by the forest industries, where one key condition is that a negative turnout of a biorefinery investment should not jeopardize the whole company. Hence, the current profitability of traditional forest-based products may be a necessary condition for a transition to a bioeconomy.

Global forest removal amounts to 3 billion m^3, of which the share of fuelwood used for cooking and heating is almost 50 %. Although it is steadily rising, the global increase in industrial wood use is still lower than the global economic growth rate and is more in pace with population growth. The importance of forests as a source of bio-based products will likely continue to grow in the future if these products replace oil-based products in some niches. The global stock of wood accounts for 527 billion cubic meters (131 m^3/ha), making up 44 % of the Earth's biomass (FAO 2010, pp. 184–185), and the world's storage of carbon in forests is an estimated 296 Gt in both above- and below-ground biomass. Over the past 25 years, the carbon stocks in forest biomass have however decreased by 697 million tons per year.

Biomass is a primary source of food, fodder and fiber, and as a renewable energy source, it provided approximately 10.2 % of the global total primary energy supply in 2008. From the IPCC expert review of available scientific literature, potential deployment levels of biomass for energy by 2050 could be in the range of 100–300 EJ. The upper bound of the technical potential of biomass for energy may be as large as 500 EJ/year by 2050. (IPCC 2011), where forest resources' contribution could reach 110 EJ (IPCC 2011, p. 228).

The potential increased use of industrial wood is difficult to estimate because it depends greatly on economic and technological developments and on unknown factors. Industrial cuttings could increase, although "there are potential implications for biodiversity, food security, and other services" (IPCC 2014, p. 816). An increasing share of raw wood materials comes from plantation forests, whereas the harvests from natural forests are decreasing (FAO 2015). The relative importance of the traditional forest products segment for the world economy is slowly diminishing, where some product groups are quite stable and others (e.g., newsprint) are in decline.

However, forest products compete intensely in the market with fossil-based products (for energy and plastics), concrete, and steel (for construction) and information technology (in the case of newsprint). In addition, the global forest sector struggles with sustainability and health issues. Traditional forest uses are however growing incrementally. Sawn wood production increased by 7 % in the 1994–2013 period. The comparable figure for paper products was 48 % over the same period, although the global production of printing paper dropped by 5 % between 2005 and 2013 (and drastically more in North America). For comparison, steel production in the past two decades (1994–2013) increased by 127 % and oil production by 25 % (World Steel Association 2015; Food and Agriculture Organization of the United Nations 2015). Hence, the current figures for the forest sector in the world economy do not reflect a booming trend in bio-based products.

The trend shift in forest product demands away from printing paper has created concerns and soul searching in the forest industries (Hurmekoski and Hetemäki 2013).

In this situation, an alternative large-volume product may be energy. Based on IPCC data, Buongiorno et al. (2012) analyzed one scenario for wood use with a considerably increased need for biomass energy and 5.5 times higher global wood energy demand. This scenario would, according to the authors, lead to increased roundwood prices and a decreased use of wood for pulp and construction. World roundwood consumption would reach 11.2 billion m^3 in 2060 and exceed sustainable harvesting levels. However, the authors claim that with a more moderate and stable fuelwood demand, the corresponding harvesting levels would only be 3.6 billion m^3, which can be combined with sustainability criteria. Additionally, world forest harvests will increasingly come from planted forests, whereas harvests from natural forests peaked in 1989. However, the key question concerning woody biomass use "…is not the amount of resources, but rather their price" (Lauri et al. 2014). The sustainability of forest resource use may naturally be difficult to assess. However, the share of forestland under certification is increasing gradually, reaching 10 % of the global forest area for the first time (UNECE/FAO 2013, p. 17, 19).

The job-creating capacity of the forest sector is gradually decreasing as larger, less labor intensive production units are being built and productivity is improving both in forest operations and in industrial processing. Employment in the forest sector shows a decreasing trend from 16 million in 1990 to less than 13 million in 2006, representing only 0.4 % of the total labor force globally. The sector's contribution to global GDP is approximately 1 %, and this share also appears to be decreasing, although the production volume of the forest sector is stable or slightly increasing (FAO 2008, p. 41; FAO 2014, p. 20).

Even though, in official charts, the forest sector plays a modest role in the world economy, it nevertheless generates informal income opportunities corresponding to 41 million additional jobs, particularly in developing countries (UNEP 2011, p. 8; FAO 2014, p. 18). Even more jobs are generated in wood-based workshops, e.g., for furniture and decoration, that is, the "IKEAs" of several rapidly growing cities in developing countries. People living in forest areas in developing countries more frequently belong to the poorest and most marginalized groups; therefore, even if the absolute monetary value of forest products is low, these products can constitute a crucial source of revenue and/or livelihood for many households and a safety net in the case of calamities (Sunderlin et al. 2008; Tesfaye 2011). Estimates of socioeconomic benefits of forests are imprecise, but a rich pool of case studies shows that non-timber forest income is of key importance for rural livelihoods in developing countries by providing products, e.g., food, honey and beeswax, and other non-wood products, calculated at $18.5 billion globally in 2005 (FAO 2010, p. 100). Other non-wood values from forests in both developing and industrialized countries include genetic material, watershed regulation, climate regulation, recreation and existence values (UNEP 2011, p. 159).

The world's forests may not yet live up to their potential to increase their contribution to a sustainable bioeconomy—in the productive or sustainability sense (FAO 2010, pp. 184–185). Deforestation is still progressing (FAO 2010, p. xxix), especially in poor countries where corruption is widespread and ecosystem services are often compromised (Shvidenko et al. 2005, p. 607; TEEB 2010, p. 16). Illegal logging is widespread, reducing opportunities to use forest revenues to enrich local economies.

Can a boosted global wood supply concord with sustainability requirements and cost restrictions? The wood-producing capacity can, in theory, be enhanced through improved silvicultural methods, tree breeding, genetically modified trees and fertilization, as changed rotation ages, fertilization and species selection increased forest production by 19 % in mid-northern Sweden (Poudel et al. 2012). However, such interventions may conflict with environmental interests and public sentiment toward genetic engineering or they can clash with other land uses. A scramble for productive land for forest production may also call for various regulations and counter-measures, including zoning, protection, and forest and agricultural intensification (Lambin and Meyfroidt 2011).

The traditional sectors of sawn wood, panels and pulp and paper will constitute the backbone of the forest sector's role in the bioeconomy for some time. Development in the forest sector is normally however piecemeal. Increased productivity and product diversification—also mundane daily improvements—are critical for a successful contribution by the forest sector to the bioeconomy. Many technologies currently available can significantly and positively impact forestry operations, for example, digital terrain models and geographical positioning systems for harvest planning, reduced ground damage and improved productivity, laser scanning for timber calculations and satellite image analysis for inventories. These technologies offer opportunities for both more productive and agile forest-based value chains, costs savings and enhanced product quality. At the same time, adoption often means a continued exodus of forest-based work opportunities from forested, rural areas.

The demand for sawn wood has decreased in the building sector in the important European and North American markets (Price Waterhouse Coopers 2013, p. 9; UNECE/FAO 2013, pp. 11–12). The forest industry has become more efficient, e.g., through optimized processes (e.g., value optimization in dry sorting lines), better monitoring of processes, more efficient energy use, and improved output ratios. X-ray log scanners and 3-D Scanning can improve log-sorting procedures and allow greater value to be created in production. Additionally, sophisticated communication opportunities between logging teams and sawmills can synchronize product streams. These examples demonstrate that even a mature industry such as sawmilling is embracing highly sophisticated production tools. Nevertheless, innovations and improvements are slower and more process oriented and incremental than in other sectors. Future improvements downstream in the wood market involve new industrial building systems based, e.g., on cross-laminated timber with the advantage of allowing tall buildings, industrial methods and better documentation of environmental advantages (carbon storage). Glulam represents ways toward low costs, fast building times, and desired strength properties, and is increasingly employed in large hall buildings, sports facilities and warehouses (UNECE/FAO 2013, p. 106). Industrial building approaches have been successfully applied to streamline the whole building process, while more house construction activities are localized indoors in controlled conditions instead of in wet environments (Sardén 2005, p. 21; Brege et al. 2013). Improvements are also affecting retailing and logistics, e.g., through an increased trading of forest products online.

Due to an expected continued fall in demand for paper for printing purposes, the pulp and paper industry must achieve a constant improvement of processes in parallel with a move outside the traditional pulp and paper product groups. Growth is still recorded in different paper segments in regions with strong economic growth rates, and new online trade and retailing methods remain economically important for the packaging segment. Gradual improvements in the pulp and paper industry include automation, process control, the more sophisticated use of sensors and more efficient and energy-saving production. The industry strives to improve energy efficiency and gradually improve different paper products in terms of their strength and printability and of the development of intelligent paper (Heikenfeld et al. 2011). The development in the forest industry is characterized by gradual efficiency improvement, such as the growth of dissolving-pulp, which can, in turn, be used in for detergents, foods, textiles and car parts. 3-D paper and "intelligent" paper—or paper with specific qualities—represent another diversifying trend.

Global figures indicate that the traditional forest industries are mature—and other industries and sectors in the world economy are growing more quickly. Hence, economic data do not suggest an imminent boom of the forest-based bioeconomy or the fundamentally improved competitiveness of traditional forest products (through cost reduction or innovation). Improvements are, of course, constant in the wood, pulp and paper industries, but they are largely gradual and in pace with innovation trends in other sectors.

The global changes in the forest sector feature ongoing restructuring and mergers, as well as greenfield investments in growth regions, such as South America and China. Other ambitions in the industry include further value-added and an increased customer focus, involving product and process innovations that also consider new business model. In particular, advanced segments such as hygiene and packaging in the quest for faster product development and a focus on solutions rather than on products is the trend.

3 Bioenergy

Wood has always been used for energy. It is also an energy source in the forest industry and in modern, large-scale Combine Heat and Power facilities. The use of wood for energy is both ancient and modern, consuming 50 % of the wood harvests, while only 10 % of the industrial energy supply comes from bioresources. Only 3 % of the world's liquid biofuels are however generated from biomass. Hence, bioenergy and biofuels are already an important component of the global energy mix. However, for further growth, more research and development support are needed.

In this way, bioenergy is one of the largest but least discussed renewable energy sources. Although bioenergy is used mostly for heating, it can also be used for the generation of electricity, which could prevent 1.3 Gt of CO_2 equivalents from being emitted into the atmosphere by 2050 (IEA 2012, p. 5). The use of biomass for electricity can furthermore be attained through co-firing with coal in combined heat and power plants. Gasification and subsequent electricity production represent yet another technology under development.

Being essential for daily needs, affordable wood for fuel also creates pollution with ensuing serious health consequences (Fullerton et al. 2008; IEA 2012, p. 14; Gordon et al. 2014). Between 600 and 800 million families, particularly women and children, are exposed to air pollution from plant-based fuels or charcoal, causing respiratory infections, pneumonia, asthma and lung cancer, which killed 3.5-4 million people globally in 2010 (Gordon et al. 2014).

In some European countries, wood energy is a key component of energy policies and goals, such as the EU 2020 goal, which stated that renewable energy sources should account for 20 % of the gross energy consumption by 2020. Through policies and economic incentives, several countries in Europe have increased the use of wood residues for energy and heat. The next stage involves increasing the use of liquid biofuels to the target of 10 % by 2020. Although first-generation biofuels (i.e., biodiesel, bio-esters, bioethanol, and biogas) have reached 3 % of total road transport fuel globally (IEA 2011, p. 11), they suffer from drawbacks because of the competition for farmland and both sustainability and economic issues—perhaps with the exception of sugarcane-based ethanol (Sims et al. 2010). Second-generation biofuels are based on a much more abundant biomass, such as lignocellulosic feedstock, and can potentially offer more sustainable and climate-neutral processes. Scale and efficiency improvements will hopefully reduce these biofuel production costs over time (IEA 2011, p. 31). For these biofuels to become truly competitive, capital requirements and operating costs must be reduced through scaling up and learning. Under this condition, biofuels can, together with substantial efficiency improvements, contribute significantly to a constant level of CO_2 emissions from the transport sector despite a large increase in vehicle fleets in developing countries (IEA 2011, p. 43). Industries are currently investing heavily in biofuel programs (Sims et al. 2010), and globally, biofuels could provide 27 % of total transport fuels by 2050, avoiding another 2.1 Gt of CO_2 emissions per year (IEA 2011, p. 21).

Much of the traditional use of wood and residues for cooking and heating is of low efficiency, causes negative health effects and is not especially sustainable. Hence, new efforts should address these issues to ultimately make this large sector of domestic bioenergy more attractive. Other improvements include the processing and upgrading of bioenergy carriers, such as drying, palletization and briquetting, which transform the material to a more energy dense and homogeneous fuel resource, and torrefaction into tar and even more energy-rich pellets and pyrolysis up to 400–600 °C can generate energy fuels with an even higher energy density (IEA 2012, p. 13).

In conclusion, bioenergy development is affected by several driving forces. Policy measures such as carbon taxes, certificates and emission trading, together with other support, have created a window of opportunity for bioenergy in different countries. Climate discussions and commitments could exercise additional forces that, in turn, will improve the competitiveness of wood energy. The current environmental and political risks associated with fossil fuel could also create advantages for biofuels, which could be both more local and more dispersed and therefore less associated with geopolitical risk (IEA-Bioenergy 2012).

However, these aspects transform the bioenergy sector into a sector with a completely different "logic" than that of the forest sector. Whereas forest products industries are operating in a global competitive environment with relatively minor trade barriers and regulations, the energy sector's bottom line is—to an important degree—dependent on policies and economic policy incentives.

4 Chemicals and Biomaterials

Annual oil production is approximately 3.9 billion tons; 92 % is used for energy and 8 % for chemicals. The share of biochemicals is 3–4 % (2010), but this share is expected to grow to 7–17 % in 2025 and further after that year (Kircher 2014). Therefore, given the ambition to create an innovative forest sector in the bioeconomy, there are opportunities to diversify and develop new bio-based chemical products. In addition to regular by-products in the pulp and paper process and biofuels, bio-based chemicals include different product classes from bulk products, specialty and fine products, and highly expensive pharmaceuticals (Kircher 2014). Posada et al. (2013) highlight that chemicals from carbohydrates possess commercial potential across different applications: energy, fuel, solvents, antiseptics, basic compounds for specialty chemicals, pharmaceuticals, industrial chemicals with many uses, softeners, food additives, bases for producing bioplastics, plasticizers, and sweeteners. However, the set of promising compounds will constantly change and that the potential of "money generators" depend on how they can provide market shares and business prospects for investors. Interesting biorefinery products range from transportation fuels with large volumes and comparably low margins to textiles, polylactic acid—which can be used to produce plastic bottles—chemicals and pharmaceuticals.

Wood plastic composites present good growth prospects and are the basis for several construction details, e.g. furniture, and decking. The bio-component of WPCs contributes to the bioeconomy, although some varieties also include fossil-based components (UNECE/FAO 2013, p. 27). Bio-based polymers, e.g., Natural-Fiber-Reinforced Composites are projected to grow many times until the late 2010s, promoted by a reduced oil supply and environmental concerns (UNECE/FAO 2013, p. 23; Philip 2013). The reinforcement of polymers by natural fibers has been applied since the first half of the twentieth century, and "Green bio composites" represent materials in which both the matrix and the fibers are bio-based and constitute a range of "building blocks" that can be produced with highly desired properties, with or without being biodegradable. Important bio-based materials also involve PLA (polylactic acid), starch plastics, bio-based PE (Polyethylene) and PHAs (polyhydroxy alkanoates). These biomaterials can enter new, environmentally sensitive markets, such as the car and boat markets (Zini and Scandola 2011; UNECE/FAO 2013, p. 32; Mizrachi et al. 2012; Fowler et al. 2006). The development of new materials is propelled by several driving forces, such as environmental concerns, recycling directives and the search for high performance in terms of density, toughness, and biodegradability. Wood-based textile thread for clothing and specialty textiles is spun from a viscose solution made from dissolving-pulp. Due to environmental, water conservation and climate reasons and to their attractiveness, cellulosic fibers are seeing increased interest from the fashion industry (Shen et al. 2010). Dissolving-pulp can also be used in such different applications as sausage casings, tire cords and as components of fillers in pharmaceutical tablets; for example, Durapulp is a combination of cellulose and biopolymer that can be pressed to two- or three-dimensional shapes. The material resembles paper, but it has other characteristics, such as high folding strength, high tear strength, high bending stiffness, high air permeability, high dimensional stability and low water absorption, and can be adapted to suit a number of applications (Roos et al. 2014).

However, improving forest-based materials' performance and functionality is not sufficient for economic and market success. To make economic sense, these products must reach high volumes, processes must be improved through a learning process, and customer preferences must be understood (Kircher 2012b). These factors engender new challenges for the traditional forest companies to forge collaborations with downstream industries, such as the car industry or manufacturing, which embrace different business models and logics.

Strategic issues involved in the development of new products are multifold but certainly not exceptional, including a deep understanding of customers' perceived value components and of concerned downstream industries (health sector, car manufacturers, chemicals, etc.). The role of the learning processes and basic economic factors must also be considered in the innovation process.

New, innovative bio-based products are generally outcomes of biorefinery production using pyrolysis, hydrolysis, biomass gasification, or other thermal, chemical or biological processes (UNECE/FAO 2013, p. 28; Fitzpatrick 2010; Naik et al. 2010). Processing steps include biofuel choice, pretreatment, and the production of intermediary products and final products. The effectiveness of enzymatic processes hinders a more intensified production of chemicals and fuels (Menon and Rao 2012). Biorefinery processes provide various products, such as construction materials, additives, and paints, as well as fuels and energy (UNECE/FAO 2013, p. 28; Sandén and Pettersson 2014), ranging from high-value-added specialty chemicals and materials to high-volume, low-value-added energy carriers. Biorefineries ideally combine production with innovation and development (Fitzpatrick et al. 2010; Centi et al. 2011).

Different biorefinery platforms require specific biomass to produce a range of products (IEA Bioenergy 2012, pp. 6–8). The challenge for biorefineries is to overcome feedstock costs, which are higher than average oil extraction costs, and to design appropriate supply systems (Kirchner 2014; Mizrachi et al. 2012). In the fossil-based economy, crude oil, which is a rather homogeneous raw material, is processed in very large refineries close to the main markets. Due to the dispersed nature of biomass, biorefineries must be located closer to the feedstock and likely be of a lesser scale than petrochemical refineries (Kircher 2012b). Important adaptations in infrastructure are needed to create cost-efficient and coordinated supply chains (Kircher 2012a). The successful development of biorefineries involves good market assessments, a wise adaptation to existing local facilities and infrastructure, quick learning, and a proper evaluation of sustainability effects (Wellisch et al. 2010)

5 The Business Environment

Although the key considerations when assessing the business of bioproducts cover a range of aspects, they are all quite straightforward and easy to grasp, ultimately coming down to the question of whether a bioproduct or service will make good business. For our analysis, this question involves a number of factors for the individual firm to consider.

The Institutional Framework
The institutional framework includes laws and regulations, as well as other policy instruments influencing the relative competitiveness of bioproducts. To some extent, product groups must consider large-scale shifts in the policy framework that may promote or block business development based on bioproducts. These aspects may have a specific importance, for instance, concerning environmental taxes on energy or decisions about nuclear energy (which, in turn, could affect bioenergy potential).

True User Needs
User needs reflect key priorities, product usage, knowledge, values and perceptions among customers and do not always immediately impact production decisions. However, with some delay, most economic production decisions are consequences of such major trends in the end market.

Industry Structure
The structure and strategic configuration the business relations include, e.g., strategic actions, market imperfections, and distribution networks. Under this label, we can also include the conditions for personal communication and collaboration within the cluster of market people, developers and planners. Opportunities and work forms for process management may ultimately determine whether market solutions can be achieved within a reasonable amount of time.

Technological Breakthroughs
Technology provides the general framework for the development of bio-based production and can change both what is possible to produce from bioresources and the costs of production.

Policies for a sustainable bioeconomy are crucial and they must of course support the substitution of bio-based products for nonrenewable products. Energy policies, certificates, taxes and fees can reduce the competitiveness of fossil energy and support the emergence and growth of bio-based energy, as environmental externalities are internalized in production costs. Forest policies can help to promote sustainable forest management, and innovation policies can promote the commercialization of new ideas. Policies must also serve to regulate conflicts in society, e.g., between production and conservation concerns, which involves handling the conflict of an increased use of bio-based resources on one hand and further developing new products and markets on the other. Caused by new bio-based products, increased demand for wood might result in tensions within the forest sector, e.g., between the traditional forest industry and bioenergy buyers. Disputes may also occur between production forestry and conservation interests focusing on ecological goods and services. Forest management is currently regulated mainly by national policies and international commitments (Puddister et al. 2011). Because the increasing world population will engender the flow of more forest-based and agricultural products alongside more ecosystem services, appropriate reconciling policies that take into consideration how land use interacts with other sectors (e.g., for employment) and the global economy (concerning trade flows) (Lambin and Meyfroidt 2011) are needed.

Many countries have implemented bioeconomy strategies and policies aimed at decoupling economic growth from environmental degradation. However, Staffas et al. (2013) observed a striking variation in how policies are envisioned and a lack of focus regarding how the concept is defined and opinions about which policies will lead to the end result. The process of forming bioeconomy policy will certainly highlight several current conflicts, including different perceptions of wood, the urban–rural divide, market forces, etc., as expected by Richardson (2012).

The ambition to increase bioresources targets a range of uses: materials, biochemical production, biofuels and biomass for heat and power (Kircher 2012a). Hence, national policies must support investments, entrepreneurs, the development of a skilled workforce, and policies for innovation in these areas. Promoting an economically competitive bioeconomy requires that these actions be well targeted to enable the establishment of appropriate infrastructure, efficient value chains, and the right kind of private–public partnerships (Kircher 2012a). Policies for advancing the bioeconomy may involve the implementation of climate policies (e.g., carbon markets) and non-discriminating regulations on materials that are open for new applications for wood in buildings (UNECE/FAO 2013, pp. 29–30). The establishment of biorefineries is best supported by subsidies for investments and support for innovation and development (Hämäläinen et al. 2011). Nevertheless, these investments will continue to be hampered by the sheer size of capital investments.

Summarizing key documents, a number of policies emerge as key drivers to globally achieving a bio-based economy: support of research and development, the creation of startups and an innovative business environment, and the provision of the proper environment for bio-based industry, that is, stable institutions, talent and know-how, and sound economic incentives. Required policy shifts include removing fossil fuel subsidies, supporting new biotechnologies in their early stages, creating a stable policy framework, introducing bioproducts sustainability indicators linked to the products' contribution to life-cycle CO_2 emissions and sustainability impacts, promoting trade in bioproducts and helping the industry solve complex supply chain and infrastructure issues.

The bioeconomy concept generally implies sustainable resource use. However, referring to Pfau et al. (2014), sustainability is not an intrinsic property of an economy based on bioresources without qualifications. Instruments are available to measure and monitor sustainability criteria. Forest certification and due diligence measures aim to combine economic use with sustainability (UNECE/FAO 2013, p. 18, p. 105). Further options include building green systems, such as LEED and BREAM, which can improve green features (UNECE/FAO 2013, p. 22). Other criteria and standards for sustainability include Global Bioenergy Partnerships, the Roundtable on Sustainable Biofuels, the ISO, and the International Sustainability and Carbon Certification System. These systems and indicators have, however, not yet reached their potential to become transparent mechanisms that help industrial and end-customers easily make the best choices.

There is still a lack of awareness in policy documents about the balance between sustainability criteria pertaining to forests and economic aspirations for a bioeconomy. Despite advances, there remains a lack of criteria and indicators on how success in this balancing act should be measured. It takes more effort to keep the forest sector at the forefront of the transition to a bioeconomy. If a bioeconomy requires more intense forest production over vast areas, the sustainability challenge for the bioeconomy is not as uncontroversial as it may seem.

6 Assessing the Business Potential of "the Wheel of Innovation Success"

Under the current market and policy environment, several factors will determine the success of bio-based product solutions and turn the bioeconomy from a vision to a reality. Once the market is identified and the product concept chosen, a "wheel of innovation" influences the logics for establishing industrial production, which have particular significance in the forest sector. All sectors in the wheel are straightforward economic factors and may pose serious obstacles; however, they may be handled once they are identified.

6.1 Customer Value

A product must provide superior value to customers. Bio-based innovations that only match current fossil-based offerings are not sufficient. Furthermore, although carbon taxes and other economic incentives and general environmental credentials may very well provide an advantage to forest-based products, most large forest industries will never base investments entirely on policy incentives that may be removed once costs are sunk. Although bioproducts may possess a positive image, sustainability claims must be supported by strict and trusted certificates. In these cases, future environmental advantages may have potential, not least from the point of view of large industrial customers and construction companies striving to provide a positive image. However, low-impact and clean processes, as well as carbon storage, must be supported by facts and figures, not only by the assertion that wood "has a good story to tell."

Bioproducts can bring new consumer benefits that have yet to be investigated thoroughly. For instance, wood surfaces are much appreciated by consumers, and exposed wooden environments in construction should be explored and used more, as the material has a competitive advantage. The advantages of wood as a construction material (e.g., light weight and strength) are increasingly being recognized.

This application is but one example, and forest-based products and chemicals must ultimately present a decisive selling argument that does not necessarily relate to the physical product as much as to the services it can provide to the customer, whether associated with chemical properties or other functions that can be created in the product (such as antibacterial paper products for hospitals).

6.2 Economies of Scale

In many cases, biorefinery products are influenced by scale effects. The processing facility may present considerable economies of scale that would justify very large units. The raw material supply system, however, will—up to a certain size—feature diseconomies of scale due to a dispersed raw material supply. The investor and planner must therefore make wise and balanced decisions to achieve economically successful units. Moreover, large greenfield investments will become prohibitively expensive. This situation can be compared with the initial phases of the current mega-companies in the Internet industry that began their learning process on modest scales and expanded their operations while experience was developed.

External economies of scale involve investments in infrastructure and resources that benefit the whole industry. For bioproducts, such investments include existing road networks, education facilities, and transporters that can accommodate large forest-based biomass resources. For comparison, the oil-based industry already has an established base of infrastructure, transportation vessels and pipelines. However, there is also a option for the bio-sector to tap into this gigantic petrochemical system and deliver bio-based chemicals to already existing petrochemical refineries. This process offers two advantages: large volume market demand together with a lesser need for facility investments.

6.3 Learning

The learning curve is an important factor for the success of innovative bioproducts. Fine-tuning and improving efficiency and a great deal of "learning by running" are real concerns in large process industries. New bioproducts may suffer from a cost disadvantage due to their feedstock base and collection systems. This relation will, however, improve as volumes are produced over time. This learning curve has already passed in oil-based refineries, whereas new competences must be developed for biorefinery operations, e.g., handling more heterogeneous raw material, determining output levels, administrating a customer base, and fine-tuning logistics, among others.

6.4 Standards and Externalities

Standards and network externalities will make the market penetration of bioproducts more complex. Standards define, for instance, how quantity and quality are measured and declared, which are absolute necessities for the creation of an efficient market and commercial activities with low transaction costs. The term "positive network externality" describes a situation in which the utility of a product for one user increases as the usage of the product widens. When the number and volume of bioproducts increase, the whole user base will be better off to the point where competition becomes too fierce. When, for instance, the use of biofuels increases, the quality and market base for vehicles adapted for this fuel increase, and the network of pump stations will also expand. This can, in turn, persuade additional customers to switch to the new fuel.

6.5 Economies of Scope

Economies of scope represent a final basic economic concept influencing the biorefinery logic. The whole meaning is to find the optimal mix of products that optimizes the refinery's economic performance. This challenge does not provide any straightforward answers. Too few products will probably mean that business opportunities from certain potential products are foregone. However, opting for too wide a range of products will make the whole production system too complex to handle because the development of each product stream may require a minimum amount of resources and separate supply chains.

Economic incentives influence the forest sector's contribution to a sustainable and an economically sound bioeconomy. Significant investments in forest biomass capacity will continue until 2050, although cost differences between fossil energy and biomass energy will remain a challenge (IEA 2012, p. 5). The largest bioenergy cost is represented by feedstock for conventional fuels (45–70 %), although for advanced fuels, capital costs will become increasingly important. It is therefore crucial to make use of by-products and extract the full economic value of coproduction processes in biorefineries.

6.6 Reality Check

Is a sustainable bioeconomy realistic and feasible on a global scale? An increasing global population—concentrated mainly in cities in developing countries (UNDESA 2014) that will become increasingly integrated in the world economy—will inevitably increase the demand for bio-based products. More sustainable and efficient forest product value chains can help meet this population's increasing needs, while simultaneously moving away from fossil fuel use. However, these new opportunities do not yet materialize in substantially increasing demand for forest products or increasing biomass capacity. The forest industry is instead being squeezed by low economic growth in industrialized countries, high timber prices, slow technology transformations, overcapacity, and low profitability (PricewaterhouseCoopers 2013, p. 5).

Customer behavior could also potentially support an emerging bioeconomy. This, in turn, warrants new, high-quality bio-based products and a knowledge and awareness about among producers about customer needs. The promotion of environmental customer choices highlights a dire need for environmental performance metrics and certification systems. According to the Ecolabel Index (2015), 458 ecolabels currently co-exist, confusing buyers and rendering decision-making and the handling of certificates complicated and costly (Räty et al. 2012, p. 42).

The challenges for biochemicals and biofuels are associated with production and handling costs, slightly more complicated processing steps, transport and logistics. Production must also become streamlined and strategically appropriate for the intended end products and supply system. A wise yet versatile combination of different output products must be conceived, which will require new collaborations and partnerships between different industry sectors. A successful business based on forest-based products requires integration with the conventional forest sector and a range of very specific competences.

6.7 Activity-Based Costing of Bioproducts

As production decisions on forest-based biorefinery products are based on delicate assessments of the profitability of each product stream, the activity-based costing (ABC) approach offers a suitable procedure for deciding on the product mix. The purpose of ABC is to calculate the cost of different products by identifying resource use and the corresponding cost drivers.

The first stage is to identify the different activities in biorefinery, e.g., procurement, handling, processing and deliveries. Resource use in each of these activities is then assessed and distributed through cost drivers of the different output products. This approach allows firms to make profitability analyses and to ultimately find the optimal product mix, which optimizes economic performance. In biorefineries, the ABC method helps to find the optimal mix of products and to ensure that costs are distributed fairly. The method is optimal for assessing biorefinery production.

7 Conclusion

The forest-based bioeconomy has promising potential. However, it is inevitable that success will not materialize in every product group. The key point is that offerings should be associated with genuine advantages for the customer. In order to create these advantages, the sector must identify real customer needs and find cost-effective (and, of course, sustainable) ways to meet these needs. The "gaps" in this process are listed below:

- Policy documents do not sufficiently consider aspects of market launch. To succeed, products must provide customer value at a reasonable cost.
- More must be learned about customer needs in both business and consumer markets. The forest industry lacks the tradition of developing offerings in companionship with important customers. New and innovative approaches to idea generation and product development should be tested.
- Market introduction, that is, learning about the market and industry, successively cutting costs, and establishing business relations, takes time and calls for stamina in business endeavors and for a willingness to learn from mistakes.
- Bioproducts are entered on competitive markets. Incumbents based on alternative raw materials must be analyzed in depth before they can be challenged.
- Basic economic fundamentals, including economies of scale, economies of scope, standards, and network externalities must be taken into consideration.
- Environmental performance of bio-products must be documented and communicated in a trustworthy and reliable way.

References

Brege S, Stehn L, Nord T (2013) Integrated design and production of multi-storey timber frame houses production effects caused by customer-oriented design. Int J Prod Econ 77:259–269

Buongiorno J, Zhu S, Raunikar R, Prestemon J P (2012) Outlook to 2060 for world forests and forest industries: a technical document supporting the Forest Service 2010 RPA assessment U S Department of Agriculture Forest Service Southern Research Station Techical Report SRS-151 Asheville NC

Centi G, Lanzafame P, Perathoner S (2011) Analysis of the alternative routes in the catalytic transformation of lignocellulosic materials. Catal Today 167:14–30

Duchesne LC, Wetzel S (2003) The bioeconomy and the forestry sector: changing markets and new opportunities. For Chronicle 79(5):860–864

EC (2012) Innovating for sustainable growth: a bioeconomy for Europe. European Commission Brussels 13, Feb 2012 COM(2012)

EC (2013) A new EU Forest Strategy: for forests and the forest-based sector. European Commission Brussels

Ecolabel Index (2015). http://www.ecolabelindex.com/. Accessed 11/5/2015

EFSOS (2010) The European forest sector outlook study II 2010–2030. UNECE/FAO, Geneva

Enriquez-Cabot J (1998) Genomics and the worlds economy. Sci Mag 281:925–926

EU Horizon2020 (2015) Horizon 2020: the EU framework programme for research and innovation. https://ec.europa.eu/programmes/horizon2020/. Accessed 29 Nov 2015

FAO (2008) Contribution of the forestry sector to national economies *1990–2006*. Food and Agriculture Organization of the United Nations, Working paper: FSFM/ACC/08. Rome 180 p

FAO (2010) Global forest assessment 2010. Food and Agriculture Organization of the United Nations Forestry paper: 163 Rome 340 p

FAO (2014) State of the World's Forests Enhancing the socioeconomic benefits from forests. Food and Agriculture Organization of the United Nations (FAO) Rome 133 p

FAO (2015) FAOSTAT http://faostat3.fao.org/download/F/FO/E. Accessed 11/5/2015. Food and Agriculture Organization of the United Nations, Rome

Fitzpatrick M, Champagne P, Cunningham MF, Whitney RA (2010) A biorefinery processing perspective: treatment of lignocellulosic materials for the production of value-added products. Bioresour Technol 101:8915–8922

Forest industries Finland (2013) Sustainable growth from bioeconomy—The Finnish Bioeconomy Strategy. Helsinki

Fowler PA, Hughes JM, Elias RM (2006) Biocomposites: technology environmental credentials and market forces. J Sci Food Agric 86(12):1781–1789

Fullerton DF, Bruce N, Gordon SB (2008) Indoor air pollution from biomass fuel smoke is a major health concern in the developing world. Trans R Soc Trop Med Hyg 102:843–851

FTP (2015) Horizons—vision 2030 for the European forest-based sector. http://www.forestplatform.org/files/FTP_Vision_revision/FTP_renewed_Vision_2030.pdf. Accessed 11/5/2015

Glesinger E (1949) The coming age of wood. Simon and Schuster, NY

Goldstein IS (1978) Chemicals from wood. Paper presented at the eighth World Forestry Congress, Djakarta

Gordon SB, Bruce NG, Grigg J, Hibberd PL, Kurmi OP, Lam KH, Mortimer K, Asante KP, Balakrishnan K, Balmes J, Bar-Zeev N, Bates MN, Breysse PN, Buist S, Chen Z, Havens D, Jack D, Jindal S, Kan H, Mehta S, Moschovis P, Naeher L, Patel A, Perez-Padilla R, Pope P, Rylance J, Semple S, Martin WJ (2014) Respiratory risks from household air pollution in low and middle income countries. Lancet Respir Med 2014:1–34

Hall R, Smolkers R, Ernsting A, Lovera S, Alvarez I (2012) Bioeconomy versus biodiversity. Global Forest Coalition. 18 http://globalforestcoalition.org/wp-content/uploads/2012/04/Bioecono-vs-biodiv-report-with-frontage-FINAL.pdf. 31/10/14

Hardy RWF (2002) The bio-based economy. In: Janick J, Whipkey A (eds) Trends in new crops and new uses. ASHS Press, Alexandria VA

Heikenfeld J, Drzaic P, Yeo J-S, Koch T (2011) A critical review of the present and future prospects for electronic paper. J SID 19(2):129–156

Hurmekoski E, Hetemäki L (2013) Studying the future of the forest sector: review and implications for long-term outlook studies. For Policy Econ 34:17–29

Hämäläinen S, Näyhä A, Pesonen H-L (2011) Forest biorefineries—A business opportunity for the Finnish forest cluster. J Cleaner Prod 19:1884–1891

IEA (2011) Technology roadmap—biofuels for transport. International Energy Agency, Paris, France 62 p

IEA (2012) Technology roadmap—bioenergy for heat and power. International Energy Agency, Paris, France 62 p

IEA-Bioenergy (2012) Bio-based chemicals—value added products from biorefineries. International Energy Agency Task 42 Biorefinery, Paris, France, 34 p

IPCC (2011) Renewable energy sources and climate change mitigation. In: Edenhofer O et al (eds) Special report of the intergovernmental panel on climate change. Cambridge University Press, Cambridge

IPCC (2014) Climate Change 2014: Mitigation of Climate Change. In: Edenhofer O, Pichs-Madruga R, Sokona Y, Farahani E, Kadner S, Seyboth K, Adler A, Baum I, Brunner S, Eickemeier P, Kriemann B, Savolainen J, Schlömer S, von Stechow C, Zwickel T, Minx JC (eds) Contribution of working group III to the fifth assessment report of the intergovernmental panel on climate change. Cambridge University Press, Cambridge

Kircher M (2012a) The transition to a bioeconomy: emerging from the oil age. Biofuels Bioprod Biorefin Biofpr 6(4):369–375

Kircher M (2012b) The transition to a bioeconomy: national perspectives. Biofuels Bioprod Biorefin Biofpr 6(3):240–245

Kircher M (2014) The emerging bioeconomy: industrial drivers global impact and international strategies. Ind Biotechnol 10:11–18

Lambin EF, Meyfroidt P (2011) Global land use change economic globalization and the looming land scarcity. Proc Natl Acad Sci 108(9):3465–3472

Lauri P, Havlík P, Kindermann G, Forsell N, Böttcher H, Obersteiner M (2014) Woody biomass energy potential in 2050. Energy Policy 66:19–31

Menon V, Rao M (2012) Trends in bioconversion of lignocellulose: biofuels platform chemicals and biorefinery concept. Prog Energy Combust Sci 37:522–550

Mizrachi E, Mansfield SD, Myburg AA (2012) Cellulose factories: advancing bioenergy production from forest trees. New Phytol 194(1):54–62

Naik SN, Goud VV, Rout PK, Dalai AK (2010) Production of first and second generation biofuels: a comprehensive review. Renew Sustain Energy Rev 14:578–597

OECD (2011) Towards green growth. Organisation for economic co-operation and development, Paris, 144 p

Pfau SF, Hagens JE, Dankbaar B, Smits AJM (2014) Visions of sustainability in bioeconomy research. Sustainability 2014(6):1222–1249

Poudel BC, Sathre R, Bergh J, Gustavsson L, Lundström A, Hyvönen R (2012) Potential effects of intensive forestry on biomass production and total carbon balance in north-central Sweden. Environ Sci Policy 15:106–124

Philip J (2013) OECD policies for bioplastics in the context of a bioeconomy 2013. Ind Biotechnol 19–21

Posada JA, Patel AD, Roes A, Blok K, Faaij APC, Pate MK (2013) Potential of bioethanol as a chemical building block for biorefineries: preliminary sustainability assessment of 12 bioethanol-based products. Bioresour Technol 135:490–499

PricewaterhouseCoopers (2013) Global forest paper & packaging industry survey 2013 edition—survey of 2012 results, Ontario, Canada

Puddister D, Dominy SWJ, Baker JA, Morris DM, Maure J, Rice JA, Jones TA, Majumdar I, Hazlett PW, Titus BD, Fleming RL, Wetzel S (2011) Opportunities and challenges for Ontario's forest bioeconomy. Forestry Chronicle 87(4):468–477

Pülzl H, Kleinschmit D, Arts B (2014) Bioeconomy—an emerging meta-discourse affecting forest discourses? Scand J For Res 9(4):386–395

Richardson B (2012) From a fossil-fuel to a bio-based economy: the politics of industrial biotechnology. Environ Plann C-Government Policy 30(2):282–296

Roos A, Lindström M, Heuts L, Hylander N, Lind E, Nielsen C (2014) Innovation diffusion of new wood-based materials—reducing the "time to market". Scand J For Res 29(4):394–401

Räty T, Lindqvist D, Nuutinen T, Nyrud AQ, Perttula S, Riala M, Roos A, Tellnes L GF, Toppinen A, Wang L (2012) Communicating the environmental performance of wood products. In: Working papers of the Finnish Forest Research Institute, 230 Vantaa, Finland

Sandén B, Pettersson K (eds) (2014) Systems perspectives on biorefineries 2014. Chalmers University of Technology, Göteborg, Sweden. ISBN 978-91-980973-2-0-

Sardén Y (2005) Complexity and learning in wood frame housing—the case of a solid wood pilot project, PhD thesis, LTU-DT-2005: 43, Luleå University of Technology, Luleå

Shvidenko A, Barber C V, Persson R (2005) Forest Systems. Ch 21. In: Rashid Hassan R, Scholes NA (eds) Ecosystems and human well-being: current state and trends, vol 1. Millennium Ecosystem Assessment Series. Island Press, Washington DC

Shen L, Worrell E, Patel M (2010) Environmental impact assessment of man-made cellulose fibres. Resour Conserv Recycl 55(2):260–274

Sims REH, Mabee W, Saddler JN, Taylor M (2010) An overview of second generation biofuel technologies. Bioresour Technol 101:1570–1580

Staffas L, Gustavsson M, McCormick K (2013) Strategies and policies for the bioeconomy and bio-based economy: an analysis of official national approaches. Sustainability 5:2751–2769

Sunderlin WD, Dewi S, Puntodewo A, Muller D, Angelsen A, Epprecht M (2008) Why forests are important for global poverty alleviation: a spatial explanation. Ecol Soc 13(2):Art 24

Swedish Forest Industries Federation (2012) The forest industry—the driver for a sustainable bioeconomy—Sustainability Report 2012, Stockholm

TEEB (2010) The economics of ecosystems and biodiversity: mainstreaming the economics of nature: a synthesis of the approach conclusions and recommendations of TEEB. The economics of ecosystems and biodiversity (TEEB). Progress Press, Malta

Tesfaye Y (2011) Participatory forest management for sustainable livelihoods in the Bale Mountains Southern Ethiopia. PhD thesis. Acta Universitatis agriculturae, Sueciae Swedish University of Agricultural Sciences, Uppsala, Sweden

United Nations (2012) The future we want. United Nation Conference on Sustainable Development, Rio de Janeiro, Brazil

UNECE/FAO (2013) Forest products annual market review. United Nations Economic Commission for Europe/Food and Agriculture Organization of the United Nations, Geneva

UNDESA (2014) World Urbanization Prospects 2014 revision. United Nation Department of Economic and Social Affairs, New York, United States

UNEP (2011) Forests—investing in natural capital. United Nations Environment Programme 193 p

White House (2012) National bioeconomy blueprint. Washington, DC, 43 p

Wellisch M, Jungmeier G, Karbowski A, Patel MK, Rogulska M (2010) Biorefinery systems—potential contributors to sustainable innovation. Biofuels Bioprod Biorefining-Biofpr 4(3):275–286

World Steel Association (2015) Statistics. https://www.worldsteel.org/. Accessed 11 May 2015

Zini E, Scandola M (2011) Green composites: an overview. Polym Compos 32(12):1905–1915

Responding to the Bioeconomy: Business Model Innovation in the Forest Sector

Eric Hansen

Abstract Myriad forces are causing fundamental changes in the operating environment of the forest sector as well as in the individual firms striving to compete in twenty-first century, globalized markets. The growing bioeconomy presents a major opportunity for forest sector firms to develop new products, explore new markets, and develop new business models. Innovativeness differs across an industry's life cycle. Product innovation is high at the beginning of an industry's life cycle and diminishes over time. Process innovation is initially low, peaks later in the life cycle, and eventually tails off to a level similar to product innovation. Forest sector companies have a traditional business culture and operate in a highly mature industry where process innovation and high-volume production tend to be the focus. The resulting culture and context of the industry does not place it well for transition to the bioeconomy. This chapter explores these issues including consideration of pathways for forest sector firms to pursue in order to capitalize on the growing bioeconomy.

Keywords Innovation · Business model · Bioeconomy · Ambidexterity

1 Introduction

Myriad forces are causing fundamental changes in the operating environment of the forest sector as well as in the individual firms striving to compete in twenty-first century, globalized markets. A key force that the industry has long dealt with is the pressure to reduce environmental impacts of operations and forest sector companies have become increasingly adept at managing and mitigating environmental impacts. Recent decades have seen changes in the environmental issues of primary concern. In the 1970s, the focus was on emissions to water and air, by the 1980s and 1990s it

E. Hansen (✉)
Department of Wood Science and Engineering, College of Forestry,
Oregon State University, 119 Richardson Hall, Corvallis, OR 97331, USA
e-mail: eric.hansen@oregonstate.edu

A. Kutnar and S.S. Muthu (eds.), *Environmental Impacts of Traditional and Innovative Forest-based Bioproducts*, Environmental Footprints and Eco-design of Products and Processes, DOI 10.1007/978-981-10-0655-5_7

was deforestation, and presently it is climate change. The interesting dynamic with respect to climate change and forest sector companies is that in many ways companies are now, at least partially, on the other side of the issue. In other words, the forest management practices of the industry, as well as the renewable materials it produces, provide positive contributions to the environment. The strategies and tactics needed to capitalize on the opportunity presented by the bioeconomy, an economy relying on biological rather than fossil resources (Roos 2016), is in sharp contrast to those needed for mitigation of negative impacts.

Today we are witnessing the infancy of the bioeconomy and there is much yet to materialize with respect to how the forest industry will embrace the opportunity (Roos and Stendahl 2015). As purveyors of renewable materials, which are increasingly seen in a favorable light, forest sector companies should be in a position to capitalize on current marketplace trends and some suggest that companies "must" transform to enter these new markets (Chambost et al. 2009). However, forest sector companies have a traditional business culture and operate in a highly mature industry where process innovation, high-volume production, and low costs tend to be the focus (Bull et al. 2015; Pätäri et al. 2011; Chambost et al. 2008). Insufficient R&D and a general unwillingness to change are said to plague parts of the industry (Hämäläinen et al. 2011) often translating to poor financial performance (Neale et al. 2014; Pätäri et al. 2011).

So, how are forest sector companies attempting to enter the new market space offered by the bioeconomy and how successful are their early efforts? This chapter explores the fundamental challenges that established companies in a traditional industry sector face as they contemplate diversifying into new products and new markets. The innovation required to successfully make this transition is not necessarily a strength of the sector, or for that matter, any existing company since, "… innovation is difficult for well-established companies. By and large, they are better executors than innovators, and most succeed less through game-changing creativity than by optimizing their existing businesses (de Jong et al. 2015, p. 1)." While optimization is necessary, especially in a sector where the wood as raw material represents such a high proportion of total costs, it is unlikely enough to successfully take full advantage of the growing bioeconomy, as we will see in subsequent pages.

2 The Opportunity of the Bioeconomy

The bioeconomy has been sufficiently explained earlier in this book, so there is no need to attempt for further detail here. The important point is that development and growth of the bioeconomy presents an opportunity for forest sector firms to diversify product offerings and move away from heavy reliance on stagnant markets for mature products. The opportunities are partially in traditional market spaces such as solid and engineered structural wood products for housing or nonresidential construction. However, the buzz is primarily around products that are relatively new to the sector, liquid fuels, chemicals, bioplastics, nanocellulose, etc. (Cai et al. 2014).

Nanocellulose, for example, presents the opportunity for "wood" to be used in products such as computer screens, films and filters, and even in computer chips (Hogan 2015). The common theme with these bioproducts is their expected/perceived superior environmental profile compared to alternatives, especially those based on petroleum. Arguably, all forest sector products are bioproducts and therefore the companies producing them are already at the heart of the bioeconomy. However, the social and political changes that are driving growth of the bioeconomy give forest-based, renewable materials a new edge in traditional markets and new opportunities across a broad host of potential new products/markets.

Especially for companies in the most developed economies, the bioeconomy may be an opportunity to develop vitality that has been lacking in the recent years, especially after the global financial crisis. As outlined previously, positioning a traditional company to capitalize on the opportunity of the bioeconomy is a significant challenge, but the alternative may be a gradual decline as new uses for forest fiber increase demand (Söderholm and Lundmark 2009) and drive up the cost of wood raw materials. Firms that fail to embrace the opportunity of the bioeconomy may find it increasingly difficult to maintain profitable operations.

Today there are numerous policies and programs coming from multiple levels of government designed to grow the bioeconomy. Think of the impacts of EU energy policies on the use of biomass for energy. Combine these with national policies in a country like Sweden and on the market side of the equation it can mean as much as 50 % of energy coming from renewable sources (SEA 2015). On the supply side, it basically has meant creation of an entirely new global industry producing wood pellets for the European market. The fast-paced development of pellet production is a striking example of the potential effects of policies on markets. However, pellets are essentially a low-value use for the wood raw material and one must question the bottom line environmental impacts of large scale, global trade in wood pellets. Does it really make environmental sense to manufacture wood pellets in the US South and ship them across the Atlantic to EU markets trying to meet their renewable energy commitments?

There are also policies in countries specifically designed to move the forest sector more quickly into the bioeconomy (Window 1). Programs such as that in Canada are essentially subsidizing investments, thereby reducing risks faced by companies.

Window 1: Government Investment in Moving an Industry to the Bioeconomy

In 2010 the Canadian federal government instituted a program called Investments in Forest Industry Transformation with a goal to stimulate innovation and create a more commercially and environmentally sustainable forest sector. As described in its first progress report, the program, "…builds the potential for new pathways into the emerging bio-economy…". With an original budget of $100C million, the program is described as highly successful in creating business transformation and bringing the next wave of innovation to market. Investments in the program have thus come far in four broad categories: biochemical, bioenergy, biomaterials, and solid wood. Overall, the estimated ROI for the projects is 25.3 % with an average payback period of 4.5 years. This is a prime example of how policy can positively impact the ability of industry to innovate and make deeper connections with the bioeconomy.

Governments are also playing an important role through R&D funding. For example, the Northwest Advanced Renewables Alliance is a US Department of Agriculture funded project seeking to produce aviation biofuel from forest residuals. The project was funded at $40 million and there are a number of other biomass-focused projects across the US representing similarly large investments. While these projects continue, entrants from outside the forest sector are making investments in biofuel production. As an example, red rock biofuels recently announced a $200 million investment in rural Southern Oregon in the US to supply biofuel for the airline industry. The company has signed supply agreements with Southwest Airlines and FedEx (Proctor 2015). Although a forest sector company will be an important supplier, red rock comes from outside the forest sector and is thus taking market share that could be taken by a forest sector company. On the other hand, only time will tell if red rock is making a good investment.

With respect to the forest sector and the bioeconomy, forest-based biorefineries have received the most attention and are described as having the biggest potential because of the logical role of pulp and paper companies and integration of the biorefinery concept to their existing large-scale operations (Hämäläinen et al. 2011) (Window 2). There is a huge range of fuels and chemicals possible from a wood-based biorefinery. As an example of the potential for a forest sector company, Lego recently announced a $150 million R&D investment with the express purpose of eliminating petroleum-based plastics in its famous bricks. The goal is to develop a biomaterial substitute by 2030 (Trangbaek 2015). Imagine a forest sector company that becomes an exclusive supplier to Lego!

Window 2: Global Forest Sector Companies and the Bioeconomy

A basic search for "bio" and "chemical" within the annual reports of the top five PricewaterhouseCoopers (Neale et al. 2014) global pulp, paper, and packaging companies, shows the focus of these companies on what can be described as "new bioproducts."

International Paper (IP): The only references to "bio" in the 2014 annual report are with respect to risk and tax factors but have nothing to do with products. It is noted in the report that under the leadership of the past CEO, the chemical division of the company was sold. Today, the company divides its sales among three broad product categories: industrial packaging, consumer packaging, and printing papers. It appears that the company has no special focus on new bioproducts.

Kimberly-Clark (KC): The 2013 annual report provides no coverage with respect to bio or chemical other than chemicals that serve as raw materials for company products. The company has four broad product categories: personal care, consumer tissue, K-C professional, and health care. KC appears to place little emphasis on new bioproducts.

Stora Enso (SE): Stora Enso begins its 2014 annual report with, "Stora Enso (the Group or the Company) is a leading provider of renewable solutions in packaging, biomaterials, wood and paper on global markets." SE has a biomaterials' division with sales of over one billion dollars (includes pulp production). In June of 2014, SE purchased the US company, Virdia, Inc. and later that year announced investment in a demonstration and market development facility in Louisiana for industrial validation of Virdia's technology. The acquisition of Virdia is described as supporting, "…the vision of Stora Enso's Biomaterials division in becoming a significant player in biochemicals and biomaterials." The company has the following broad product categories: consumer board, packaging solutions, biomaterials, wood products, paper, etc. New bioproducts are clearly an important element of SE's focus.

Svenska Cellulosa (SCA): SCA is committed to tripling its production of biofuels from forests by 2020, with 2010 as the reference year. However, when using the term "biofuel", the company refers to forest-based residuals and peat—not liquid fuels. SCA has the following broad product categories: personal care, tissue, and forest products. SCA appears to have no special focus on new bioproducts.

UPM-Kymmene (UPM): UPM refers to itself as The Biofore Company. The UPM's 2014 annual report outlines the new business development for long-term growth in three areas, biofuels, biochemical, and biocomposites. The company began producing wood-based diesel in early 2015 in its Finland operations which it claims as a world's first. UPM's vision reads as follows: "As the frontrunner of the new forest industry UPM leads the integration of bio and forest industries into a new, sustainable and innovation-driven future…" UPM operates with six business areas: UPM Biorefining, UPM Energy, UPM Raflatac, UPM Paper Asia, UPM Paper Europe and North America, and UPM Plywood. Wood sourcing and forestry, biocomposites, and biochemicals are in "other operations." UPM clearly has the strongest new bioproducts focus of the five largest global pulp, paper, and packaging firms.

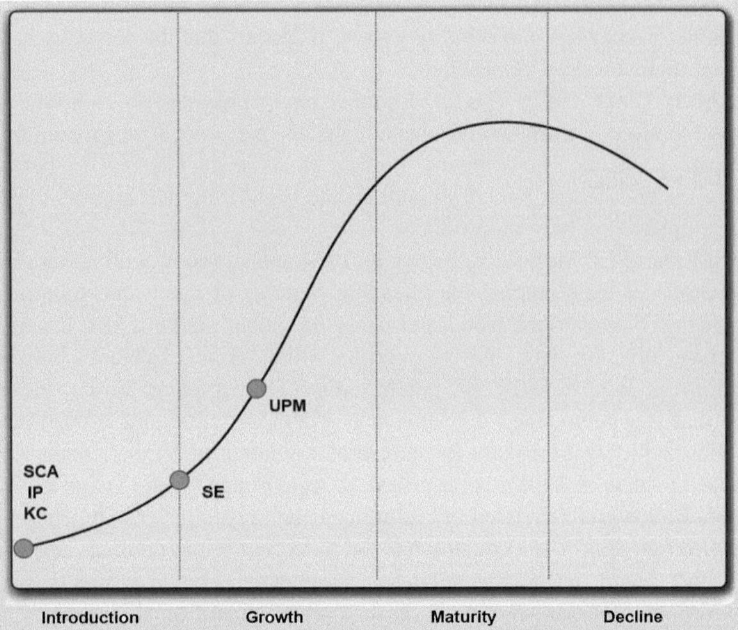

Global forest sector companies and their place in the "new" bioproducts lifecycle

The accompanying graphic shows UPM-Kymmene as the early adopter of the five companies, followed by Stora Enso. The question remains, will early adoption pay off?

Beyond biorefineries, the bioeconomy presents a broad suite of opportunities for companies in all sectors of the industry. Tall wood buildings are currently a major phenomenon in the global forest sector and cross-laminated timber is the product of choice for such buildings. New products, designs, and/or adaptations of the basic cross-laminated timber concept will likely follow. Following the pattern in middle Europe, companies producing glulam are logical candidates to enter this area of the bioeconomy. Beyond tall buildings, housing applications in general need to be made more sustainable and wood-based products may be important contributors to the process. Accompanying cross-laminated timber and other building products is the need for more environment friendly adhesives, connectors, and other supporting materials and products. Society is increasingly demanding natural, chemical free, and local products and wood products are often well-suited to meet these demands. Incremental steps towards the bioeconomy can be made by traditional manufacturers by creating, for example, environmental product declarations supporting the green nature of their products and better fitting the paradigm of the bioeconomy. Unique new uses from wood or other renewable materials will increasingly fit evolving marketplace demands for environmentally friendly products (Window 3).

Window 3: Examples of Bioeconomy Upstarts

Onbone Oy, a company in Helsinki, Finland, produces medical products from wood and biodegradable plastic. The founders of the company have their roots in the Chemistry Department at the University of Helsinki. The company's primary product is cast material (Woodcast®) for orthopedic and traumatology uses. The company claims Woodcast to be "naturally better" and describes the product as, "…a completely non-toxic and perfectly mouldable and remouldable cast material suitable for all casting work. It is manufactured from clean wood and biodegradable plastic and can be molded without water or rubber gloves." As a replacement for gypsum-based plaster, this renewable material is a prime example of what might be offered by a bioeconomy company. http://www.woodcast.fi/en/company.

Ecovative is a US material science company based in New York that produces renewable products from Mushroom® Materials. The company began with packaging and has since branched into surfboards, automotive, insulation, and other applications. The basic idea is to grow biocomposites. Mycelium is combined with agricultural wastes such as stalks and seed husks in a customized mold for the desired packaging. The mycelium grow on and around the agricultural waste, thus forming a low-density packing element the shape of the mold. When the final consumer receives her product she can simply compost the accompanying packaging. Thus, Ecovative is providing an environmentally friendly alternative to petroleum-based packaging materials and provides another illustrative example of a bioeconomy company. http://www.ecovativedesign.com/.

American Process Inc. is a US company based in Georgia providing nanocellulose fibrils and crystals under the brand BioPlus™. According to company materials, "BioPlus can be used to make lighter and more fuel-efficient cars and planes, stronger and more streamlined building materials and cements, and more natural personal care products and cosmetics." It also claims tremendous environmental and sustainability advantages over petroleum-based products. The proprietary manufacturing process used by American Process allows it to produce nanocellulose products from forest and agriculture residues, bamboo, and grasses. Using this broad range of renewable raw materials is a key aspect of this bioeconomy company. https://americanprocess.com/bioplus/.

Generally, any improvements to make products more environment friendly are steps toward the bioeconomy. There are many existing examples of this such as various forms of environmental certification, nonformaldehyde-added adhesives, and reduced density fiberboards. Environmental product declarations are a means for the industry to formally document the performance of its products and can provide a means for measuring improved performance. More radical steps in this direction would be developing new products with a design for environment philosophy, thereby minimizing environmental impacts over the entire lifecycle of the product. In this way, forest sector products can play a more significant role in the circular economy where products are actively reused, repaired, refurbished and recycled (Window 4).

When people think of recycling, they think of paper to paper or aluminum to aluminum, but there are other ways of reuse. Upcycling takes a used product or material and turns it into something of higher value. Think of pallets being made into furniture. The design community can play an important role in envisioning opportunities for upcycling. The Italian company "alcarol" uses transparent substances to preserve materials in their original form. The company's bricola collection uses pilings from Venice and its marble ways line uses wood slabs from the marble cutting process. Both are used in the production of high-end furniture (alcoral 2015).

Window 4: Wood Products in the Circular Economy

CaReWood: Cascading Recovered Wood is an EU research project designed to, "introduce an upgrading concept for recovered solid timber as a source of clean and reliable secondary wooden products for the European industry." The project began in 2014 and is scheduled for a 2017 completion. It includes 15 partners from five different countries. CaReWood has both technical and economic aspects. For example, one element is to provide forecasts of the volumes and qualities of postconsumer and postindustrial recovered wood. Guidelines are being developed for such things as "design for disassembly" and software supported reverse logistics models. Certification and labeling criteria will be developed along with an evaluation of social and environmental impacts.

The bioeconomy presents an opportunity for forest sector firms to round out their product portfolios to include new, growth products to accompany their existing mature products. A more balanced portfolio should help with profitability issues that have been prevalent in the sector in recent years. However, as has been outlined earlier, many firms in the sector may not be especially well-equipped to make this move. For companies to make the leap to the bioeconomy, what is needed? There have been many calls for increased innovation within the forest sector.

3 Innovation

Innovation is the buzz word that invariably rises when discussing a more competitive future forest sector and the importance of innovation to the future of the sector has become well-accepted (Björkdahl and Börjesson 2011). Myriad governmental programs in recent years have been designed to develop a more innovative industry. Examples include Canada's Investments in Forest Industry Transformation program (see Window 3), the EU's Forest-based Technology Platform, New Zealand's Growing Confidence in Forestry's Future Research Programme, and Australia's ForestWorks training program. Despite these efforts the industry continues to be criticized for its lack of innovation (Bull et al. 2015).

What is innovation and what does it mean for an industry or a company to be innovative? An innovation is a new product, manufacturing process, or way of managing the business (Hovgaard and Hansen 2004). Small changes are often referred to as incremental innovations and something truly new-to-the-world is referred to as radical innovation. Innovativeness is a related but somewhat different concept. Innovativeness is a characteristic of an organization or individual that has the propensity to create and/or adopt new products, manufacturing processes, or business management methods (Knowles et al. 2008). When the literature describes the forest sector as mature, focused on costs, and resistant to change, these are characteristics opposite of innovativeness. If forest sector companies are to remain viable competitors with substitute materials and global rivals, new products, processes, and business systems are critical. In other words, innovativeness and out-of-the-box thinking are called for across the sector (Hämäläinen et al. 2011; Hansen et al. 2007). One reason the forest sector lacks innovation is that it is a mature industry and is, therefore, well into its innovation life cycle.

3.1 The Innovation Life Cycle

The level of innovation intensity within an industry sector tends to differ based on the general evolution of that sector. Innovation tends to be higher during early stages of the life cycle and declines as industries mature (Utterback 1994). Product innovation is generally higher in early stages while process innovation grows in importance as product innovation declines (Window 5).

In the beginning of an industry there is significant experimentation with product design among competitors. Thus, the rate of product innovation is high while little concentration is placed on process innovation. Once a general or dominant design is accepted in the marketplace, the focus shifts to innovation in manufacturing processes. Later in the life cycle companies are focused on costs, volume, and capacity. At this stage, product and process innovation occurs in small incremental steps (Utterback 1994) and this evolution continues until some external shock (technological change, new competition, etc.) (Tushman and Nadler 1986) occurs that brings a new wave of product innovation. Utterback's (1994) theory suggests that as companies producing mature products see the imperative for change, they dedicate increased resources to new product development (NPD). A contemporary example of this is Levi Strauss Company. At 162 years old, the company has been producing the same basic product for many years, blue jeans. However, the last two decades have seen a fall in sales of over $2 billion and most recently a new entrant, yoga pants, is a key reason for falling sales. A new CEO was hired in 2011 to rejuvenate the company. This has meant significant R&D and a new focus on women's jeans (Higgins 2015).

Window 5: Evolution of Industry Innovation

Utterback (1994) uses the auto industry to illustrate how innovation changes over time as shown in the accompanying graphic. In the very early stages of the auto industry the car took many forms. Similarly, various media were created prior to VHS becoming the accepted standard in the market. Once the dominant design is reached, the innovation focus changes to efficient production of that dominant design. Think of the Model-T Ford. As an industry matures, the overall level of each type of innovation decreases. According to the theory, as a highly mature industry, the forest sector is on the extreme right side of the accompanying graphic. New products and/or new markets are needed to start a new lifecycle.

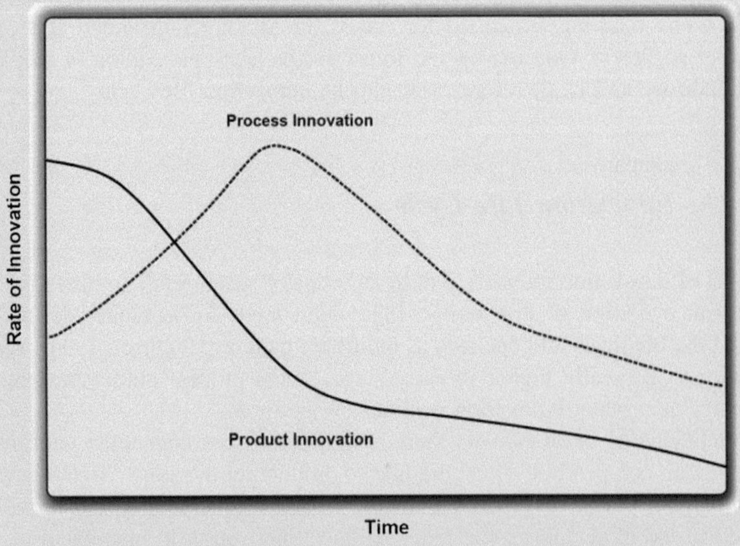

A pattern of incremental change followed by major discontinuities is common across industry sectors. Examples of discontinuities include new technologies, competitors, regulations, and even recessions (Tushman and O'Reilly 1996). For the forest sector, the following examples are noteworthy:

- Information technology—newsprint
- The rise of China—furniture and hardwood plywood
- Illegal logging legislation—entire sector

- The Global Financial Crisis—most firms, but especially North American structural building products manufacturers

Within a specific product category, competition from substitutes can be a key discontinuity. In outdoor decking, preservatively treated wood decking substituted for naturally durable species. This was followed by wood-plastic composites (WPCs) gaining major market share from treated wood. Today, inroads are being made by companies offering modified wood decking.

Firms that survived the global financial crisis are likely quite well-placed for continued near-term success. But, are they now positioning themselves to survive the next discontinuity? Forest sector firms have long relied on the merits of wood being a renewable material and thus highly environment friendly. What happens when a bioeconomy startup comes to market with an environmentally superior product to those based on wood? Responding to discontinuities requires one or more forms of innovation. While Utterback's (1994) work focused on product and process innovation, the fast-paced changing dynamics of modern markets means that the capability to innovate, create, and/or adopt new business models is essential.

3.2 Business Model Innovation

The way a company goes about creating and providing a valuable product and/or service to its customers is often referred to as its business model. Teece (2010) provides this description.

> The essence of the business model is in defining the manner by which the enterprise delivers value to customers, entices customer to pay for value, and convert those payments to profit. It thus reflects management's hypothesis about what customers want, how they want it, and how the enterprise can organize to best meet those needs, get paid for doing so, and make a profit.

Following this definition, business model innovation refers to creating and/or adopting new business models. More generally, innovation is possible across all aspects of managing a business and is often referred to as organizational innovation or business systems innovation.

As forest sector companies create new products or target new customers, new or at least incrementally adapted business models are required. If a sawmilling company integrates forward into the value chain to create a subsidiary that builds passive houses, it requires a new business model. The market for high energy efficient housing is at the heart of the bioeconomy. If a wood treating company begins to offer recycling of its products at end-of-life, it requires a new business model (Window 6). Recycling is also at the heart of the bioeconomy.

Forest sector companies are not well known for sophisticated marketing practices and research again shows them to be laggards in this area (Han and Hansen 2015). New products and new markets require new approaches for customer communication. For example, even for business-to-business markets, social media will grow in importance and firms must develop in this area to successfully launch new business models. This illustrates that developing new business models requires new knowledge and skills. For some companies, the innovation in this aspect of the business model may be more challenging than, for example, developing a new product.

Window 6: Cox Industries Enters New Space in Bioeconomy

Cox Industries is a family-owned company that provides a broad portfolio of treated wood products. Recognizing the mature and highly competitive nature of most of its markets, the company recently underwent a strategic innovation initiative. A chief innovation officer was hired and the company undertook a systematic evaluation of innovation opportunities. One opportunity identified was a recycling service offered to its utility company customers. Disposal of poles by, for example, electrical utilities, is a significant time and money burden as well as a significant risk factor. By offering a guaranteed 72-h disposal pickup and accessing a network of EPA-approved incineration facilities, Cox was able to create a totally new company, CoxRecovery. This diversification allowed Cox to access new space within the bioeconomy, but it required creation of a totally new business model than that used for its traditional treating business.

On the pulp and paper side of the industry, replacing petroleum products using cellulosic feedstocks via large biomass consuming operations (biorefineries) is seen by many as a key path to the bioeconomy. For biorefineries to become a reality, the industry needs information about new business models and cross-industry collaboration will likely be necessary, resulting in new networks and ways of integrating (Hämäläinen et al. 2011; Söderholm and Lundmark 2009). In fact, it is said that development of forest-based biorefineries may translate to a fundamental structural change for the paper industry (Näyhä and Pesonen 2014). In practice this translates to a switch in mentality from making paper first and selling the by-products second to making chemicals first and treating paper as the by-product. Unfortunately, research indicates that firms are not especially adept at rethinking in existing business models (Björkdahl and Börjesson 2011; Teece 2010). Biorefineries are not a totally new phenomenon and much can be learned from the experience of existing operations and the business models they have employed over time (Window 7).

Window 7: Biorefineries Not a New Concept

The biorefinery concept is far from new and some firms have operated in this market for decades. Borregaard Industries Ltd in Norway and SEKAB BioFuel Industries AB in Sweden are two prime examples. Borregaard began producting ethanol in 1938 and has since broadened its product line to a wide range of specialty chemicals. The company has renewed its business model multiple times over the years as it dealt with competition from petroleum and companies from low-cost regions. Business model renewal was based on significant R&D investments to develop new products as well as branching out to nonwood feedstocks. Today, the company derives its products from cellulose, hemicellulose, and lignin, thereby maximizing the suite of possible outputs (Rødsrud et al. 2012).

Research shows that forest sector firms are adept at innovation in manufacturing processes, but are less proactive with respect to product and business systems innovation. However, as noted above, these are the types of innovation most needed for embracing the potential of the bioeconomy. Innovativeness and the propensity to create and/or adopt innovations is embedded in the culture of the organization, so culture is a fundamental issue that can facilitate, or stand in the way of, capitalizing on the bioeconomy.

4 The Role of Culture

Culture plays a critical role in the ability to make transitions required for entry to the bioeconomy. My recent research focused on innovation within the forest sector and shows that culture is a dominant driver of innovation or, alternatively, a significant hurdle to innovation in the industry (Hansen et al. 2007). Quotes from industry managers can quickly elucidate the influence of culture. Reacting to a question about the potential for innovation in his company a US manager simply said, "We're working with wood, so what is there?" In sharp contrast, in responding to a similar question, a European manager stated, "…we are really trying to, as we say in our strategy, we are developing innovative solutions for our customers." Other researchers provide illustrative remarks from industry managers, "the biggest problem with this firm is its management. Most of the top management has grown up with big mills and established customers, which require very little innovation. All new issues and ideas are horror for them." (as quoted in Björkdah and Börjesson 2011). One can easily imagine which of these three companies is most likely to innovate.

Tushman and O'Reilly (1996) refer to cultural and structural inertia that make it difficult for companies to make larger changes. They describe structural inertia as, "…resistance to change rooted in the size, complexity, and interdependence in the organization's structures, systems, procedures, and processes (p 18)." Cultural inertia is the result of norms, values, and lessons that make up the accepted "way of doing things" within a firm. The more ingrained these norms and values become, the greater the inertia. One can just imagine a manager reacting to a new idea from an employee, "That's not how we do things around here." This exemplifies how culture can get in the way of business model renewal that may be essential for long-term firm survival (Tushman and O'Reilly 1996).

How an industry or company defines itself is a cultural issue and can significantly impact long-term success. The historic troubles of the railroad and movie industries illustrate how a narrow definition can make it more difficult to cope with discontinuities. When the railroads defined themselves as railroad companies, instead of transportation companies, it meant they did not deal well with the transition to truck transportation. Similarly, the Hollywood movie industry initially held little interest in television (Levitt 1960). It is a very different game if a forest industry company defines itself as "a forest products company" versus "a building products company" versus "a bioproducts company." Similarly, a paper producer that defines itself as "a paper company" will have a very different view of the world (and culture) than one that defines itself as a sustainable packaging company. The sustainable packaging company might see a startup like Ecovative (Window 3) with its mycelium-based packaging as a potential target for purchase while Ecovative would be unlikely to even hit the radar of the paper company.

At the risk of repeating, the common thread with respect to culture in the forest industry has been and largely remains a commodity mentality with a strong focus on low costs and efficiencies. In many ways this is logical since innovative process technology has been recognized as key to competing in mature industrial markets, and has also been shown to increase firm financial performance (Cohen and Sinclair 1990). But the longstanding focus of being production oriented and focusing on process technology does not necessarily provide companies with the tools necessary to effectively deal with competition from new, substitute products. More specifically to this chapter, it does not equip firms with the skills needed to effectively capitalize on the opportunities presented by the bioeconomy. The culture that facilitates increasing efficiencies and a constant focus on cost reduction is just the culture that can prevent developing innovative ideas and entering new markets. Although focusing on the bioeconomy is not necessary for all firms, companies looking to enter the bioeconomy need to develop an ability to maintain focus on efficiency while simultaneously actively exploring new opportunities.

4.1 Ambidexterity

A wealth of research exists showing that when a company is highly effective at decreasing costs, and increasing efficiencies (exploitation), it is not necessarily also endowed with the skills necessary to create new products, services, etc. (exploration). Being effective at both is typically referred to as organizational ambidexterity (Tushman and O'Reilly 1996). In its most basic sense, organizational ambidexterity means that a firm can do two different things simultaneously (Simsek 2009).

Imagine the outcome of a recently announced project at Umeå University where the goal is 3-D printing of cellulose-based materials for home construction (Anonymous 2015). What sort of company is most likely to embrace this technology and turn it into a commercial reality? Is a sawmilling company that is highly skilled at manufacturing a traditional wood building product or a construction company that is steeped in the current building technology and focused on creating buildings as efficiently as possible, well-placed to make the leap to 3-D printing of houses? Probably not. That is why it is often startups or companies outside an industry that typically introduce radical innovation (Window 3). Recent research shows that some experts on the paper industry feel the industry culture will not support successful development, production, and marketing of high, value-added products, because the industry has the tradition of commodities. These experts suggest that energy or oil companies will be the ones that capitalize on biorefineries. In other words, forest sector companies will provide the raw materials and the oil and energy companies will gather most of the profits (Näyhä and Pesonen 2014). If forest sector companies do not capitalize on the opportunity, companies outside the sector will fill the void (Näyhä and Pesonen 2012).

Outdoor decking in the US marketplace is an example of companies from outside the sector offering a substitute product that quickly gained market share. Wood-plastic composites (WPCs) for outdoor decking were pioneered by Trex, a company that was originally part of Mobil Chemical. Since the establishment of Trex in 1996, many other companies have entered the market with plastic and WPC decking, including some forest products companies. Whether petroleum-based plastics, even when recycled, can be considered to have a proper role in the bioeconomy is an open question, but a modified wood product that can exhibit some of the primary advantages of WPCs could be a way for the traditional decking manufacturer to capitalize on the bioeconomy and potentially increase competitiveness. Accoya® Wood and various providers of heat treated wood such as Oy Lunawood Ltd are good examples of companies offering modified wood products positioned as environmentally friendly products for the bioeconomy. So what are strategies for enhancing ambidexterity or otherwise facilitating the entry of forest sector firms into the bioeconomy?

5 Pathways to the Bioeconomy

This chapter is written with a bias that forest sector companies should be entering the bioeconomy in order to assure their long-term viability. While I readily acknowledge that this is debatable, what I argue is undeniable is the need to continuously innovate and renew products, processes and business models. Doing so to pursue the bioeconomy may not, in the final analysis, be the right path forward for every company, but the company that is innovating will be ready to respond to whatever trends that take place in the larger economy. As one reviewer of this chapter suggested, "While you are tweaking the optimizer in the mill, someone outside the sector is developing a product that will make yours obsolete!" Choosing not to proactively innovate may inadvertently turn into an exit strategy.

Having said the above, most will also agree that momentum toward the bioeconomy is also undeniable. Consumers are demanding more environmentally friendly products and mega issues such as climate change demand rethinking products and economies regardless of individual consumer demands. So, where should forest sector companies start in their attempt to enter the bioeconomy? I advocate throughout the text above that culture is the foundation from which companies can begin to innovate. Get the culture element right and the process will be much easier since instead of hindering innovation, culture will facilitate the process. Embedded in the culture should be a common understanding of the fundamental reason the company exists, are you a building products company, or a bioproducts company, for example. One cultural norm should be a spirit of close monitoring of the competition, especially potential competitors from outside the sector. Following the development of bio-products and startups using similar feedstocks is an area in which to pay special attention. This can help in preparing for defensive competitive moves or more offensive moves such as purchases of intellectual property or closely mimicking competitor offerings.

How does a tradition-bound company in a mature industry sector go about entering the bioeconomy? Incremental cultural tweaks can push a company toward enhanced innovativeness. For the average company, an increased focus on the end user (higher level of market orientation) of its products can open up many opportunities for innovation. Many firms know their customer, but not their customer's customer. New knowledge about how the product is used in its final application can help identify opportunities for product adaptations as well as process improvements. Recognizing and acting on these opportunities can lead to successes that will feed organizational change toward a more innovative culture. There is growing evidence that internal social networks are a tool that can serve to enhance innovation by increasing the interaction among employees in an organization (Rossi 2015). More aggressive efforts toward ambidexterity involve creating new structures for exploration (Birkinshaw and Gibson 2004). Several viable options include the following:

- A separate R&D unit
- Cross-functional innovation (NPD) teams
- Creating a closely held subsidiary company
- Alliance/collaboration with other companies
- Purchase of another company

At some level, each of these allow disruption of the existing cultural norms and if done well can facilitate effective exploration and innovation to complement the exploitation skills possessed by most firms in the industry.

5.1 Establish a Separate R&D Unit

A separate R&D unit can possess its own innovative culture that is distinct from the rest of the company. Accordingly, this is one structure that can increase the ability of a company to explore new things while the remainder of the company focuses on exploitation. However, many forest sector companies have downsized or even eliminated their R&D departments and in most cases, at least in the short term, these units will not be returning. For those companies without the option of creating or using a separate R&D unit, an option to facilitate exploration is to actively collaborate with research institutes and universities on relevant projects. Industry/University collaboration comes in a variety of forms. Direct funding of projects is common as is participation in research cooperatives. The US-based wood-based composites center provides a good example of industry/university collaboration that produces results (Window 8).

Collaboration with research organizations can be combined with accessing governmental programs designed to facilitate research, development, and innovation, especially in small companies. In the US, the small business innovation research program is an example. In the EU the example is Horizon 2020. The "SME Instrument" will provide approximately three billion Euros in funding between 2014 and 2020, exclusively to small and medium-sized enterprises. These kind of programs reduce the risk exposure of the participating company, often requires working with research institutes and/or universities, and can also facilitate networking with like–minded companies across the economy.

Window 8: Examples of Industry/University Collaboration

The Wood-Based Composites Center is an Industry/University Cooperative Research Center that is sponsored by the National Science Foundation. Financial support is provided by NSF and member organizations. Members include domestic and international corporations and government agencies. The Center was originally formed in 1999 and currently includes twelve member organizations and four universities: Virginia Tech, Oregon State University, University of Maine, and University of British Columbia. The mission of the Center is to, "Advance the science and technology of wood-based composite materials." Research conducted by the Center focuses on fundamental science and is therefore precommercial in nature. This allows member companies equal access to fundamental new knowledge they can apply in their own product/process development and commercialization efforts. Center research has led to new commercial developments and process improvements in the areas of adhesive formulation, adhesive application methods, adhesive additives, and composites processing (Kamke 2015). http://wbc.vt.edu/center/.

The Centre for Advanced Wood Processing at the University of British Columbia is a national center for education, training and outreach, research, product development, and technical assistance for wood products manufacturers. The Center was formed in 1997 to support the value-added manufacturing sector. The mission of the Center is, "…to offer training and technical assistance to promote the success and sustainable growth of Canada's value-added wood products manufacturing sector." Collaboration with industry is partially through applied research and training, but the cornerstone of collaboration is through the Center's cooperative education program. Students in the UBC Department of Wood Science Wood Products Processing degree program typically spend between 16 and 20 months working in industry as part of their degree. This gives the student more marketable skills upon seeking permanent employment and companies get a low-risk way of evaluating individuals as potential permanent employees. http://cawp.ubc.ca/.

5.2 *Establish Cross-functional Teams*

Another way of creating an exploration structure is to pull together innovative individuals from across the company to form innovation teams. A well-designed cross-functional team can operate outside the exploitation culture of a mature organization, essentially operating with its own subculture. Teams can collaborate with various outside organizations like the examples in the previous section. Networking outside the organization exposes the team to a broader set of ideas and can facilitate creation and/or adoption of new products, processes, and business systems. Still if the team identifies a good opportunity such as offering a new product, it still must convince the rest of the organization to give it a genuine try. Because the new product would likely mean sacrificing efficiency (exploitation), transferring the concept to the larger organization can come with significant challenges.

5.3 *Create a Subsidiary*

Creating a closely held subsidiary company that is in firm control of the existing executive team is a method identified as highly successful in research by O'Reilly and Tushman (2004). Cox Industries and CoxRecovery in Window 6 is an example of this approach. This, of course, requires an executive team with a sufficiently innovative culture to envision the creation of a new company and its business model. The new company is able to develop its own unique cultural norms without carrying the baggage of the culture in the parent company, making it is easier to explore and experiment with business models appropriate for new products and/or new markets. It also means the ability to establish a brand that is not tied to the parent company.

5.4 *Initiate Cross-company Collaboration*

Some researchers emphasize that cross-sector collaboration will be especially important, for example, in biorefineries because new knowledge and new business models will be needed (Hämäläinen et al. 2011). Of course, collaboration can take many forms ranging from joint R&D to joint ventures like those outlined in Window 9. Lego's effort to create a sustainable alternative to its petroleum-based plastic seems a ripe opportunity for a forest sector company to link up for joint R&D. It should be emphasized here that collaboration with companies with contrasting cultures can prove challenging (Window 9), yet interaction with highly different companies (in terms of products, culture, expertise, etc.) promises to be the most fruitful, again because difference leads to exposure to new ideas and concepts.

Envisioning how forest products can interact with other products via the Internet of Things is a prime avenue for cross-sector, cross-company collaboration and innovation. For example, smart packaging made by a paper company could be designed to effectively interact with smart industrial refrigeration units at the wholesale and retail levels as well as a smart refrigerator in a consumer's home. In the context of the Internet of Things, what role might structural wood products play in both residential and commercial settings? Moisture content monitoring might be the most obvious example. In the value-added arena, furniture is being fitted with touch screens that allow the user to make calls, listen to the radio, and browse the Internet (Banks 2014). Already commercialized is a line of furniture designed to monitor user data such as weight and blood pressure, with the option of delivering that data to health care providers (Tolentino 2015). Furniture giant IKEA collaborated with design firm IDEO and students at several universities to develop futuristic ideas for furniture that were presented at the 2015 Milan Design Week. Shelves that act as refrigerators and tabletops that cook were to examples presented (Briodagh 2015). Most forest sector firms will not have in-house expertise on the Internet of Things, so collaborating in this realm could open up many innovation opportunities.

Window 9: Notable Bio-focused Joint Ventures in the Forest Sector

Catchlight Energy LLC: In 2008 Weyerhaeuser and Chevron entered into a joint venture, Catchlight Energy LLC, to create technology to convert cellulosic feedstocks to biofuels. The idea was to connect the forestry expertise and infrastructure of Weyerhaeuser with Chevron's marketing and sales infrastructure (Casey 2013). Chevron claims that after investigating 50 conversion technologies and 100 feedstocks that no financially viable solution materialized. In reality, estimates of 5–10 % profits were not competitive with the 17 % average at Chevron (Elgin and Waldman 2013). According to the 2014 Weyerhaeuser annual report, Catchlight was dissolved in 2014.

NSE Biofuels Oy: In 2007 Stora Enso and Neste joined forces to create this joint venture aimed at producing biodiesel from forest residues. The companies built a demonstration plant that successfully converted biomass into bio-based diesel (Voegele 2012). The initial commercial scale plant was estimated to cost slightly over $600 million, but the companies chose not to move forward with the investment.

5.5 *Purchase Another Company*

Finally, outright purchase of another company is a way to access new products and markets (Stora Enso buying Virdia, Window 2). Indeed, it can be a way of gaining a new culture, though integrating cultures can be a major challenge as has been experienced via numerous mergers. Purchasing can allow for quick entry into new markets or new products as well as the ability to quickly match competitor offerings. The management team at the purchased company can diversify that of the parent company and this can cause the parent company to slowly evolve its culture and strengthen exploitation skills.

6 Conclusions

A growing bioeconomy means the opportunity for numerous types and forms of innovations. New products and services will be offered using a variety of business models serving an increasingly environment-focused consumer base. How forest sector firms capitalize on the opportunity is yet to be seen. As purveyors of products from renewable materials, companies in the sector are endowed with a head start, an inherent advantage, over companies with less environmentally friendly offerings. However, this head start will quickly dissipate without genuine strategies to further tailor products and services for the bioeconomy customer.

Not every company needs to develop new-to-the-world products based on cutting edge research and development. However, even the most traditional companies serving the most traditional markets should be thinking about their role in the bioeconomy and how their products should evolve to ensure long-term viability. Innovation in the bioeconomy will happen with or without current forest sector companies. To join the momentum toward the bioeconomy will require innovation in products and business models and this process is far from easy.

Facilitating innovation in traditional, process oriented companies should start with culture. Several mechanisms for enhancing innovation (exploration) are discussed above. Each mechanism has the potential to begin shifting organizational culture so that it is more exploration focused. A more balanced and ambidextrous organization is better positioned for long-term survival in the fast-changing bioeconomy.

Acknowledgements Thanks to Scott Leavengood of Oregon State University's Wood Innovation Center and Lyndall Bull of Lynea Advisory for their insight and advice while writing this chapter.

Literature

alcoral (2015) http://www.alcarol.com/. Accessed on 24 Aug 2015

Anonymous (2015) Sliperiet at Umeå University is starting a project: 3D printing houses. http://www.umu.se/english/about-umu/news-ents/news/newsdetailpage/sliperiet-at-umea-university-is-starting-a-project-3d-printing-houses.cid254168?utm_content=buffer1c13f&utm_medium=social&utm_source=twitter.com&utm_campaign=buffer. Accessed on 23 July 2015

Banks R (2014) How the Internet of things will change everything. Mobile Industry Review. http://www.mobileindustryreview.com/2014/11/internet-of-things.html. Accessed on 31 July 2015

Birkinshaw J, Gibson C (2004) Building ambidexterity into an organization. MIT Sloan Mgmt Rev 45:47–55

Björkdahl J, Börjesson S (2011) Organizational climate and capabilities for innovation: a study of nine forest-based Nordic manufacturing firms. Scan J of For Res 26(5):488–500

Briodagh K (2015) IKEA makes furniture smarter with smart kitchens. http://www.iotevolutionworld.com/smart-home/articles/402857-ikea-makes-furniture-smarter-with-smart-kitchens.htm. Accessed on 31 July 2015

Bull L, Hansen E, Jenkin B (2015) Maximising the potential of Australia's forests—collaborating and innovating to realise the opportunity. Workshop report and response to the Forest Industry Advisory Council's strategic directions issues paper. Lynea Advisory, Melbourne, Australia, 25 p

Cai Z, Rudie AW, Stark NM, Sabo RC, Ralph SA (2014) New products and product categories in the global forest sector. In: Hansen E, Panwar R, Vlosky R (eds) The global forest sector: changes, practices, and prospects. CRC Press, Boca Raton

Casey T (2013) Why Chevron won't make bio-gasoline for $2.18 any time soon. Clean Technica. http://cleantechnica.com/2013/04/23/the-catchlight-chevron-biofuel-project-stalls-out/. Accessed on 27 July 2015

Chambost V, McNutt J, Stuart PR (2008) Guided tour: implementing the forest biorefinery (FBR) at existing pulp and paper mills. Pulp Pap Can 109(7–8):19–27

Chambost V, McNutt J, Stuart PR (2009) Partnerships for successful enterprise transformation of forest industry companies implementing the forest biorefinery. Pulp Pap Can 110(5):19–24

Cohen DH, Sinclair SA (1990) The adoption of new manufacturing technologies: impact on the performance of North American producers of softwood lumber and structural panels. For Prod J 40(11–12):67–73

de Jong M, Marston N, Roth E (2015) The eight essentials of innovation. McKinsey Q. April. http://www.mckinsey.com/insights/innovation/the_eight_essentials_of_innovation. Accessed on 17 Aug 2015

Elgin B Waldman P (2013) Chevron defies California on carbon emissions. Bloomberg Business. http://www.bloomberg.com/news/articles/2013-04-18/chevron-defies-california-on-carbon-emissions. Accessed 27 July 2015

Hämäläinen S, Näyhä A, Pesonen HL (2011) Forest biorefineries–a business opportunity for the Finnish forest cluster. J of Cleaner Prod 19(16):1884–1891

Han X, Hansen E (2015) Marketing sophistication in private U.S. Sawmilling Companies. Can J For Res 46(1):181–189

Hansen E, Juslin H, Knowles C (2007) Innovativeness in the global forest products industry: exploring new insights. Can J of For Res 37(8):1324–1335

Higgins T (2015) Distressed denim. Bloomburg Bus Week, pp 48–55, 27 July 2015–2 Aug 2015

Hogan K (2015) Scientists develop functioning computer chips made out of wood. http://inhabitat.com/these-functioning-computer-chips-are-made-out-of-wood/. Accessed on 24 Aug 2015

Hovgaard A, Hansen E (2004) Innovativeness in the forest products industry. For Prod J 54(1): 26–33

Kamke F (2015) Personal communication,25 Aug 2015

Knowles C, Hansen E, Shook SR (2008) Assessing innovativeness in the North American softwood sawmilling industry using three methods. Can J of For Res 38(2):363–375

Levitt T (1960) Marketing myopia. Harvard Bus Rev 38(4):24–47

Näyhä A, Pesonen HL (2012) Diffusion of forest biorefineries in Scandinavia and North America. Technol Forecast Soc Chang 79(6):1111–1120

Näyhä A, Pesonen HL (2014) Strategic change in the forest industry towards the biorefining business. Tech Forec and Soc Chg 81:259–271

Neale D, Murdoch I, Bromley K, Henriques D (2014) Global forest, paper and packing industry survey. PricewaterhouseCoopers, 17 pp

O'Reilly CA, Tushman ML (2004) The ambidextrous organization. Harvard Bus Rev 82(4):74–83

Pätäri S, Kyläheiko K, Sandström J (2011) Opening up new strategic options in the pulp and paper industry: case biorefineries. For Policy and Econ 13(6):456–464

Proctor C (2015) Biofuel from new $200M Southern Oregon plant destined for FedEx jet engines. Portland Bus J http://www.bizjournals.com/portland/blog/sbo/2015/07/red-rock-biofuels-lands-another-big-customer-for.html. Accessed on 10 Aug 2015

Rødsrud G, Lersch M, Sjöde A (2012) History and future of world's most advanced biorefinery in operation. Biomass Bioenergy 46:46–59

Roos A (2016) The bio-economy and the forest business community. In: Kutnar A, Muthu S (eds) Environmental impacts of traditional and innovative forest-based bioproducts. Springer, Berlin

Roos A, Stendahl M (2015) The role of the forest sector in the emerging bio-economy. In: Panwar R, Hansen E, Kozak R (eds) Forests, business and sustainability. Taylor & Francis Books, London

Rossi B (2015) The evolving role of internal social networks in the workplace. Inf Age. http://www.information-age.com/it-management/strategy-and-innovation/123459336/evolving-role-internal-social-networks-workplace. Accessed on 31 Aug 2015

Simsek Z (2009) Organizational ambidexterity: towards a multilevel understanding. J Manag Stud 46(4):597–624

Söderholm P, Lundmark R (2009) Forest-based biorefineries. For Prod J 59(1/2):7

Swedish Energy Agency (SEA) (2015) Energy in Sweden—facts and figures 2012. https://www.energimyndigheten.se/Global/Engelska/Facts%20and%20figures/Energy%20in%20Sweden%20facts%20and%20figures%202012%20%282%29.pdf. Access on 31 July 2015

Teece DJ (2010) Business models, business strategy and innovation. Long Range Plan 43(2):172–194

Tolentino M (2015) New furniture tracks your health, from weight to blood pressure. Silicon Angle. http://siliconangle.com/blog/2015/04/23/new-furniture-tracks-your-health-from-weight-to-blood-pressure/. Accessed on 31 July 2015

Trangbaek RR (2015) Lego group to invest 1 billion DKK boosting search for sustainable materials. http://www.lego.com/en-us/aboutus/news-room/2015/june/sustainable-materials-centre. Accessed on 31 July 2015

Tushman M, Nadler D (1986) Organizing for innovation. Ca Mgmt Rev 28(3):74–92

Tushman ML, O'Reilly CA III (1996) Ambidextrous organizations: managing evolutionary and revolutionary change. Cal Mgmt Rev 38(4):8–30

Utterback JM (1994) Mastering the dynamics of innovation, how companies can seize opportunities in the face of technological change. Harvard Business School Press, Boston, 253 pp

Voegele E (2012) Neste oil, Stora Enso shelve plans for renewable diesel plant. http://biomassmagazine.com/articles/7962/neste-oil-stora-enso-shelve-plans-for-renewable-diesel-plant. Accessed on 30 July 2015

MIX
Papier aus verantwortungsvollen Quellen
Paper from responsible sources
FSC® C105338

If you have any concerns about our products,
you can contact us on
ProductSafety@springernature.com

In case Publisher is established outside the EU,
the EU authorized representative is:
Springer Nature Customer Service Center GmbH
Europaplatz 3, 69115 Heidelberg, Germany

Printed by Libri Plureos GmbH
in Hamburg, Germany